REVIEWS in MINERALOGY

(Formerly: "Short Course Notes")

Volume 4

MINERALOGY and GEOLOGY of NATURAL ZEOLITES

FREDERICK A. MUMPTON, Editor

The Authors

James R. Boles
  Department of Geosciences
  University of California
  Santa Barbara, California 93106

Edith M. Flanigen
  Linde Division
  Union Carbide Corporation
  Tarrytown, New York 10591

Arthur J. Gude, 3rd
Richard A. Sheppard
  U.S. Geological Survey
  Federal Center
  Denver, Colorado 80225

Richard L. Hay
  Department of Geology & Geophysics
  University of California
  Berkeley, California 94720

Frederick A. Mumpton
  Department of the Earth Sciences
  State University College
  Brockport, New York 14420

Ronald C. Surdam
  Department of Geology
  University of Wyoming
  Laramie, Wyoming 82070

Series Editor

Paul H. Ribbe

Department of Geological Sciences
Virginia Polytechnic Institute & State University
Blacksburg, Virginia 24061

MINERALOGICAL SOCIETY OF AMERICA

COPYRIGHT

PRINTED BY

BookCrafters, Inc.
Chelsea, Michigan 48118

REVIEWS IN MINERALOGY

(Formerly: SHORT COURSE NOTES)

ISSN 0275-0279

VOLUME 4: MINERALOGY and GEOLOGY
of NATURAL ZEOLITES

ISBN 0-939950-04-9

ADDITIONAL COPIES

Additional copies of this volume as well as those
listed below may be obtained at moderate cost from

Mineralogical Society of America
2000 Florida Avenue, NW
Washington, D.C.   20009

Vol. 1: SULFIDE MINERALOGY, P.H. Ribbe, Editor (1974)   284 p.

Vol. 2: FELDSPAR MINERALOGY, P.H. Ribbe, Editor (1975; revised 1981) ~300 p.

Vol. 3: OXIDE MINERALS, Douglas Rumble III, Editor (1976)   502 p.

Vol. 4: MINERALOGY and GEOLOGY of NATURAL ZEOLITES,
         F.A. Mumpton, Editor (1977)   232 p.

Vol. 5: ORTHOSILICATES, P.H. Ribbe, Editor (1980)   381 p.

Vol. 6: MARINE MINERALS, Roger G. Burns, Editor (1979)   380 p.

Vol. 7: PYROXENES, C.T. Prewitt, Editor (1980)   525 p.

Vol. 8: KINETICS of GEOCHEMICAL PROCESSES, A.C. Lasaga and
         R.J. Kirkpatrick, Editors (1981)   391 p.

Vol. 9A: AMPHIBOLES and Other Hydrous Pyriboles - Mineralogy,
         D.R. Veblen, Editor (1981)   372 p

Vol. 9B: AMPHIBOLES: Petrology and Experimental Phase Relations,
         D.R. Veblen and P.H. Ribbe, Editors (1981)   ~375 p

REVIEWS in MINERALOGY

(Formerly: "Short Course Notes")

FOREWORD

In 1974 the Mineralogical Society of America began publishing paperback books in conjunction with short courses presented at the time of its annual meetings. The purpose of these courses and their accompanying "Short Course Notes" has been to assemble cogent and concise reviews of the literature and current research in particular disciplines or subject-areas in mineralogy, petrology, and crystallography.

In 1980 the Council of the Mineralogical Society of America changed the title of the serial publications proceeding from its short courses from "Short Course Notes" to "Reviews in Mineralogy" in order to reflect more accurately the content of these volumes. The reviews are basically designed as 'reference textbooks' for postgraduates and are used as such in numerous academic, government, and industrial institutions. Titles of all volumes published through 1981 are listed on page ii; they are available at moderate cost from the Mineralogical Society of America.

This volume, *Mineralogy and Geology of Natural Zeolites*, was originally published in 1977 in conjunction with a short course sponsored by M.S.A. and convened by F.A. Mumpton of the State University of New York, College at Brockport, New York. This is its second printing in a slightly revised and updated version. A notable addition is Appendix III, *X-ray Powder Diffraction Patterns of Common Zeolites in Sedimentary Rocks* by Arthur J. Gude, 3rd.

Paul H. Ribbe
Series Editor
Blacksburg, VA
June 30, 1981

MINERALOGY and GEOLOGY
of NATURAL ZEOLITES

Preface and Acknowledgments

This volume was prepared to serve as notes for a short course on the Mineralogy and Geology of Natural Zeolites held in Seattle, Washington, November 4-6, 1977. Sponsored by the Mineralogical Society of America, the short course was organized by the following individuals who served as lecturers, authors, and staff:

J. R. Boles	F. A. Mumpton
E. M. Flanigen	R. A. Sheppard
A. J. Gude, 3rd	R. J. Stewart
R. L. Hay	R. C. Surdam

Although several excellent reviews of the geological occurrences, the mineralogical properties, and the industrial and agricultural applications of natural zeolites have appeared in the recent literature, the authors hope that these notes will be of value not only to the participants in the short course itself, but to others in the geological and mining community who wish to become better acquainted with the modern ideas about zeolite minerals.

The title of the short course reads "Natural Zeolites"; however, the subject matter treated in the course and reviewed here deals primarily with those zeolites that occur in sedimentary rocks and which have formed by authigenic or burial diagenetic processes. Unfortunately, only limited coverage has been given to the classical occurrences of zeolites--the megascopic crystals in the vugs and cavities of basalts and other basic igneous rocks. Our only justification is that since the late 1950s, almost all major efforts on zeolites have been directed towards the "sedimentary" occurrences, and it is these occurrences of zeolites in sedimentary rocks that are still unfamiliar to many geologists and mineralogists. It is our intention that this short course and these notes will play a small role in alleviating this unfamiliarity.

Special thanks are due to Paul Ribbe and Ms. Margie Strickler for invaluable assistance during the preparation of the notes for publication and to Douglas Rumble for his many words of advice during the organization of the short course. Technical assistance was graciously provided by the Department of the Earth Sciences, State University College, Brockport, New York, the Department of Geosciences, University of Washington, and Mr. John L. Gannon, Portland, Oregon. Our appreciation is also expressed to Pergamon Press, Inc. and Wiley-Interscience for their permission to reproduce much of the material in several of the chapters and to other publishers for use of countless tables and illustrations.

Frederick A. Mumpton
Brockport, New York
November 1977

Table of Contents

	Page
Copyright and Additional Copies	ii
Foreword	iii
Preface and Acknowledgments	iv
Recommended References	x

Chapter 1   NATURAL ZEOLITES   F. A. Mumpton

INTRODUCTION	1
MINERALOGY	2
Chemical Composition	2
Crystal Structure	4
HISTORICAL	8
REFERENCES	15

Chapter 2   CRYSTAL STRUCTURE AND CHEMISTRY OF NATURAL ZEOLITES   Edith M. Flanigen

INTRODUCTION	19
CRYSTAL STRUCTURES	20
Group 1 (S4R)	22
Analcime	22
Phillipsite	28
Gismondine	28
Laumontite	31
Group 2 (S6R)	31
Erionite	31
Offretite	34
Sodalite	34
Group 3 (D4R)	34
Zeolite A	34
Group 4 (D6R)	37
Faujasite and Faujasite-type Structures	37
Chabazite	41
Group 5 (T_5O_{10} Units)	41
Natrolite, Scolecite, Mesolite	41
Thomsonite, Gonnardite, Edingtonite	41
Group 6 (T_8O_{16} Units)	45
Mordenite	45
Dachiardite, Ferrierite	48

Chapter 2 (continued)

    Group 7 ($T_{10}O_{20}$ Units)   48
        Heulandite, Clinoptilolite   48
  ACKNOWLEDGMENTS   52
  REFERENCES   52

Chapter 3   GEOLOGY OF ZEOLITES IN SEDIMENTARY ROCKS   R. L. Hay

  INTRODUCTION   53
    Mineralogy   53
    Occurrences   55
  SALINE, ALKALINE LAKES DEPOSITS   57
  SOILS AND SURFACE DEPOSITS   58
  DEEP-SEA DEPOSITS   58
  OPEN-SYSTEM TYPE DEPOSITS   60
  HYDROTHERMAL DEPOSITS   60
  BURIAL DIAGENETIC DEPOSITS   61
  ZEOLITES IN IGNEOUS ROCKS   62
  REFERENCES   63

Chapter 4   ZEOLITES IN CLOSED HYDROLOGIC SYSTEMS   Ronald C. Surdam

  INTRODUCTION   65
  GEOLOGIC SETTING   66
    Modern Closed Basins   66
    Hydrographic Features   67
    Brine Evolution   68
    Weathering Reactions   69
    Chemical Trends   70
    Efflorescent Crusts   73
    Volcanism   74
  MINERAL PATTERNS   74
    Pleistocene Lake Tecopa Deposits   77
    Pliocene Big Sandy Formation   77
    Miocene Barstow Formation and Eocene Green River Formation   79

Chapter 4 (continued)

    INTERPRETATION OF MINERAL PATTERNS 80
        Formation of Alkalic, Silicic Zeolites from Silicic Glass 80
        Reaction of Alkalic, Silicic Zeolites to Analcime 82
        Reaction of Zeolites to Potassium Feldspar 84
        Hydrochemistry 86
    REFERENCES 89

Chapter 5   ZEOLITES IN OPEN HYDROLOGIC SYSTEMS   R. L. Hay and
                                                                           R. A. Sheppard

    INTRODUCTION 93
    ZEOLITE FORMATION IN SILICIC TEPHRA DEPOSITS 94
    ZEOLITE FORMATION IN LOW-SILICA TEPHRA DEPOSITS 99
    REFERENCES 102

Chapter 6   ZEOLITES IN LOW-GRADE METAMORPHIC ROCKS   James R. Boles

    INTRODUCTION 103
    SELECTED ZEOLITE OCCURRENCES FROM THE CIRCUM-PACIFIC AREA 105
        New Zealand 105
        Japan 107
        Western North America 112
            British Columbia - Triassic rocks 112
            British Columbia - Cretaceous rocks 113
            British Columbia - Cretaceous to Eocene rocks 113
            Washington to Alaska - Tertiary rocks 113
            Washington - Tertiary rocks 115
            Oregon - Triassic and Jurassic rocks 115
            California - Cretaceous rocks 115
            California - Tertiary rocks 116
    LITHOLOGY, CHEMISTRY, AND STABILITY OF ZEOLITE SPECIES 117
        Heulandite-group Minerals 117
            Lithology 117
            Chemistry 117
            Stability 119
        Analcime 120
            Lithology 120
            Chemistry 120
            Stability 120

Chapter 6 (continued)

    Laumontite 122

        Lithology 122
        Chemistry 124
        Stability 124

    Less Common Zeolites 124

        Wairakite 124
        Mordenite 126
        Stilbite 126
        Other zeolites 126

  ZEOLITE REACTIONS 126

    Effect of Pressure 128

    Effect of Temperature 128

    Effect of Parent Material and Bulk Composition 129

    Effect of Fluid Chemistry 130

    Effect of Permeability 131

    Effect of Kinetics 132

  REFERENCES 132

Chapter 7   ZEOLITES IN DEEP-SEA SEDIMENTS   James R. Boles

  INTRODUCTION 137

  OCCURRENCE AND MINERALOGY OF PHILLIPSITE AND CLINOPTILOLITE 138

    Age of Sediment 138

    Type of Sediments 138

    Depth of Burial 141

    Areal Distribution 142

    Associated Authigenic Minerals 142

    Mineralogy 142

        Phillipsite 142
        Clinoptilolite 147

    Origin of Phillipsite and Clinoptilolite 151

  OTHER ZEOLITES IN DEEP-SEA SEDIMENTS 155

    Analcime 155

    Harmotone 157

    Laumontite 158

    Erionite 158

    Zeolites in Volcanic Rocks 158

Chapter 7 (continued)

COMPARISON OF DEEP-SEA ZEOLITE OCCURRENCES WITH OTHER OCCURRENCES 159
Conditions of Formation 159
Distribution of Zeolites 159
Mode of Zeolite Occurrences 159
REFERENCES 160

Chapter 8   COMMERCIAL PROPERTIES OF NATURAL ZEOLITES   E. M. Flanigen
                                                                    and F. A. Mumpton

INTRODUCTION 165
ADSORPTION PROPERTIES 165
CATION EXCHANGE PROPERTIES 168
DEHYDRATION AND DEHYDROXYLATION PHENOMENA 171
REFERENCES 174

Chapter 9   UTILIZATION OF NATURAL ZEOLITES   F. A. Mumpton

INTRODUCTION 177
POLLUTION CONTROL APPLICATIONS 177
Radioactive Waste Disposal 178
Sewage Effluent Treatment 180
Agricultural Wastewater Treatment 182
Stack-Gas Cleanup 182
Oil-Spill Cleanup 183
Oxygen Production 183
ENERGY-CONSERVATION APPLICATIONS 184
Coal Gasification 184
Natural-Gas Purification 185
Solar Energy Use 187
Petroleum Production 187
AGRICULTURAL APPLICATIONS 188
Fertilizers and Soil Amendments 188
Pesticides, Fungicides, Herbicides 189
Heavy Metal Traps 189
Animal Nutrition 190
Excrement Treatment 192
Aquacultural Uses 193

Chapter 9 (continued)   Page

MINING AND METALLURGY APPLICATIONS 195
    Exploration Aids 195
    Metallurgical Uses 195
MISCELLANEOUS APPLICATIONS 196
    Paper Products 196
    Construction Uses 197
    Medical Applications 197
ACKNOWLEDGMENTS 198
REFERENCES 198

Appendix I.   REPRESENTATIONS AND MODELS OF ZEOLITE CRYSTAL STRUCTURES 205

Appendix II. CRYSTAL STRUCTURE DATA FOR IMPORTANT ZEOLITES 207

Appendix III. X-RAY POWDER DIFFRACTION PATTERNS OF COMMON ZEOLITES IN
              SEDIMENTARY ROCKS. . . . . Arthur J. Gude, 3rd   219

Recommended References

CONFERENCE VOLUMES (listed by date of publication)

Barrer, R. M., Conf. Chairman (1968) *Molecular Sieves:* Society of Chemical Industry, London, 339 pp.

Flanigen, E. M. and Sand, L. B., Eds. (1971) *Molecular Sieve Zeolites I:* Adv. Chem. Ser. 101, Amer. Chem. Soc., Washington, D. C., 562 pp.

Flanigen, E. M. and Sand, L. B., Eds. (1971) *Molecular Sieve Zeolites II:* Adv. Chem. Ser. 102, Amer. Chem. Soc., Washington, D. C., 459 pp.

Meier, W. M. and Uytterhoeven, J. B., Eds. (1973) *Molecular Sieves:* Adv. Chem. Ser. 121, Amer. Chem. Soc., Washington, D. C., 634 pp.

Uytterhoeven, J. B., Ed. (1973) *Molecular Sieves: Proceedings of the Third International Conference on Molecular Sieves:* Leuven Univ. Press, Leuven, Belgium, 484 pp.

Katzer, J. R., Ed. (1977) *Molecular Sieves - II:* ACS Symp. Ser. 40, Amer. Chem. Soc., Washington, D. C., 732 pp.

Sand, L. B. and Mumpton, F. A., Eds. (1978) *Natural Zeolites: Occurrence, Properties, Use:* Pergamon Press, Elmsford, New York, 546 pp.

Rees, L. V. C., Ed. (1980) *Proceedings of the Fifth International Conference on Zeolites:* Heyden, London, 902 pp.

Townsend, R. P., Ed. (1980) The Properties and Applications of Zeolites: Spec. Publ. 33, The Chemical Society, London, 430 pp.

Sersale, R., Colella, C., and Aiello, R., Eds. (1981) *Recent Progress Reports and Discussion, 5th International Conference on Zeolites:* Giannini, Naples, Italy, 320 pp.

(Recommended References continued on the next page)

BOOKS AND MAJOR PAPERS

Barrer, R. M. (1978) *Zeolites and Clay Minerals as Sorbents and Molecular Sieves:* Academic Press, New York, 497 pp.

Breck, D. W. (1964) Crystalline molecular sieves: J. Chem. Educ. 41, 679-689.

_____ (1974) *Zeolite Molecular Sieves: Structure Chemistry and Use:* Wiley Interscience, New York, 771 pp.

_____ and Smith, J. V. (1959) Molecular Sieves: Scientific Amer. 200, 85-94.

Brouchek, F. E., Ed. (1979) Природные Цеолиты *(Natural Zeolites):* Metzneyeryeba Publ. House, Tbilisi, 333 pp.

Coombs, D. S., Ellis, A. J., Fyfe, W. S., and Taylor, A. M. (1959) The zeolite facies, with comments on the interpretation of hydrothermal synthesis: Geochim. Cosmochim. Acta 17, 53-107.

Hays, R. L. (1966) Zeolites and zeolite reaction in sedimentary rocks: Geol. Soc. Amer. Spec. Pap. 85, 130 pp.

Iijima, Azuma (1980) Geology of natural zeolites and zeolitic rocks: In *Proceedings of the Fifth International Conference on Zeolites:* L. V. C. Rees, Ed., Heyden, London, 103-118.

Kossovskaya, A. G., Ed. (1980) Природные Цеолиты *(Natural Zeolites):* Nauka Publ. House, Moscow, 224 pp.

Krupyennekova, A. Yu., Ed. (1980)Природные Цеолиты в Сельском Хозяйстве*Natural Zeolites in Agriculture):* Metzneyeryeba Publ. House, Tbilisi, 254 pp.

Meier, W. M. and Olson, D. H. (1978) *Atlas of Zeolite Structure Types:* Structure Comm. Int. Zeol. Assoc., Distributed by Polycrystal Book Service, Pittsburgh, PA, 99 pp.

Mumpton, F. A. (1973) Worldwide deposits and utilization of natural zeolites: Industrial Minerals, 73, 30-45.

_____ (1978) Natural zeolites: a new industrial mineral commodity: In *Natural Zeolites: Occurrence, Properties, Use:* L. B. Sand and F. A. Mumpton, Eds., Pergamon Press, Elmsford, New York, 1-27.

_____ and Ormsby, W. C. (1976) Morphology of zeolites in sedimentary rocks by scanning electron microscopy: Clays & Clay Minerals 24, 1-23.

_____ and Fishman, P. H. (1977) The application of natural zeolites in animal science and aquaculture: J. Animal Sci. 45, 1188-1203.

Olson, R. H., R. H., Breck, D. W., Sheppard, R. A., and Mumpton, F. A. (1975) Zeolites: In *Industrial Minerals and Rocks,* S. J. Lefond, Ed., Amer. Inst. Mining, Metall., Petroleum Engineers, New York, New York, 1235-1274.

Rabo, J. A., Ed. (1976) *Zeolite Chemistry and Catalysis:* Amer. Chem. Soc. Monogr. 171, 769 pp.

Senderov, E. E. and Khitarov, N. I. (1970) Цеолиты, Их Синтез и Условия Образования в Природе *(Zeolites, Their Synthesis and Conditions of formation in Nature):* Nauka Publ. House, Moscow, 283 pp.

Sheppard, R. A. (1973) Zeolites in sedimentary rocks: U. S. Geol. Surv. Prof. Pap. 820, 689-695.

Surdam, R. C. and Sheppard, R. A. (1978) Zeolites in saline, alkaline-lake deposits: In *Natural Zeolites: Occurrence, Properties, Use:* L. B. Sand and F. A. Mumpton, Eds., Pergamon Press, Elmsford, New York, 145-174.

Tsitsishvili, G. V., Andronekashbili, T. G., and Krupyennekova, A. Yu., Eds. (1977) Клиноптилолит *(Clinoptilolite):* Metzneyeryeba Publ. House, Tbilisi, 243 pp.

Chapter 1

NATURAL ZEOLITES

F. A. Mumpton

INTRODUCTION

As every student of mineralogy or geology knows, zeolites are ubiquitous constituents in the vugs and cavities of basalts and other traprock formations. Beautiful assemblages of well-formed crystals up to several inches in size are prized by mineral
collectors and adorn the mineral museums of every country. Several dozen individual
species have been identified among which are chabazite, erionite, faujasite, and mordenite whose adsorption properties rival those of several synthetic molecular sieves.
In addition to their occurrence in basic eruptive rocks and related late-stage hydrothermal environments, zeolites are recognized today to be among the most abundant and
widespread authigenic silicates in sedimentary rocks. Since their discovery in the
late 1950s as _major_ constituents of numerous volcanic tuffs in ancient saline-lake
deposits of the western United States and in thick marine tuffs of Italy and Japan,
more than 1,000 occurrences of zeolites have been reported from volcanogenic sedimentary rocks of more than 40 countries. The high purities and flat-lying nature of
the sedimentary deposits have aroused considerable commercial interest both here and
abroad, and applications based on their unique physical and chemical properties have
been developed for them in many areas of industrial and agricultural technology.

Zeolites have been known for more than 200 years, but it was not until the middle
of this century that the scientific community became generally aware of their attractive properties or of their geological significance in the genesis of tuffaceous sediments. During the last 20 years, interest in zeolites has increased at a remarkable
rate, and it is currently difficult to pick up an issue of any of the leading geological or chemical journals without finding at least one article devoted to these
materials. Most of the recent effort has centered around _synthetic_ molecular sieves,
an area that has grown into multi-million dollar businesses in several countries, or
natural zeolites as they occur in sedimentary rocks of volcanic origin. Amygdaloidal
zeolites are still the principal source of high purity specimens for x-ray structure
determinations or for sophisticated experimentation, but most of the geological and

1

industrial activity has concentrated on "sedimentary" zeolites. For this reason, the present volume deals mainly with those zeolites formed in sedimentary environments-- their mineralogy and crystal chemistry, their geologic occurrence, and their industrial utilization.

MINERALOGY

By definition, zeolites are crystalline, hydrated aluminosilicates of alkali and alkaline earth cations, having infinite, three-dimensional structures. They are further characterized by an ability to lose and gain water reversibly and to exchange constituent cations without major change of structure. Zeolites were discovered in 1756 by Freiherr Axel Fredrick Cronstedt, a Swedish mineralogist, who named them from the Greek words ζειν and λιθοσ, meaning "boiling stones," in allusion to their peculiar frothing characteristics when heated before the blowpipe. From Cronstedt's vantage point, the property of ". . . gassing and puffing up almost like borax . . . followed by melting . . . to a white glass" was sufficient to distingish zeolites as a separate type of silicate mineral. Indeed, the intumescence of zeolites on heating has been used for years by amateur and professional mineralogists alike to identify this group (see Northup, 1941; Brush and Penfield, 1926). The zeolites constitute one of the largest groups of minerals known; more than 40 distinct species have been recognized, and nearly 100 species having no natural counterparts have been synthesized in the laboratory. The potential application of both synthetic and natural zeolites stems, of course, from their fundamental chemical and physical properties, which in turn are directly related to their chemical compositions and crystal structures.

Chemical Composition

Along with quartz and feldspar minerals, zeolites are tektosilicates, that is, they consist of three-dimensional frameworks of SiO_4^{-4} tetrahedra wherein all four corner oxygen ions of each tetrahedra are shared with adjacent tetrahedra, as shown in Figure 1. This arrangement of silicate tetrahedra reduces the overall Si:O ratio to 2:1, and if each tetrahedron in the framework contains silicon as the central cation, the structure is electrically neutral, as is quartz (SiO_2). In zeolite structures, however, some of the quadrivalent silicon is replaced by trivalent aluminum, giving rise to a deficiency of positive charge. The charge is balanced by the presence of mono and divalent cations, such as Na^+, Ca^{2+}, K^+, etc., elsewhere in the structure. Thus, the empirical formula of a zeolite of the type:

2

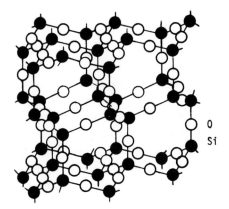

O
Si

Figure 1.   Three-dimensional arrangement of silicate tetrahedra in tektosilicates.
Crystal structure of beta-tridymite.

$$M_{2/n}O \cdot Al_2O_3 \cdot xSiO_2 \cdot yH_2O$$

where M is any alkali or alkaline earth cation, n is the valence of that cation, x is
a number from 2 to about 10, and y is a number from 2 to 7. The oxide and unit-cell
formulae of clinoptilolite, a common natural zeolite, follow:

$$(Na,K)_2O \cdot Al_2O_3 \cdot 10SiO_2 \cdot 6H_2O \quad \text{and} \quad (Na_4K_4)(Al_8Si_{40})O_{96} \cdot 24H_2O.$$

Ions within the first set of parentheses in the unit-cell formula are known as exchangeable cations; those within the second set of parentheses are called structural
cations, because with oxygen they make up the tetrahedral framework of the structure.
It should be noted that the base to alumina ratio is always equal to unity, and the
$(Al + Si):O$ ratio is always 1:2. In addition, no zeolites are known which contain
more tetrahedral aluminum ions than silicon ions; the ratio of $SiO_2:Al_2O_3$ is always
equal to or greater than 2:1.

Early mineralogical reference books and texts stressed so-called "coupled" sub-
stitutions of Na + Si for Ca + Al in zeolites. Modern ideas of ionic substitution
and cation exchange negate this concept, and, to the first approximation, cations substitute freely for one another in zeolite species, the only restriction being that of
charge balance. Thus, in a given species, $2Na^+$ can proxy for $1Ca^{2+}$, or $1NH_4^+$ for
$\frac{1}{2}Sr^{2+}$. Trivalent cations are generally not found in the exchange sites of most

3

zeolite minerals. It is possible by simple washing to produce a series of cationic forms of a single species differing only in the nature of the cations in the exchange positions. The cation composition of natural samples, therefore, reflects in large part the composition of the last solution to which they were exposed. Such phenomena pose serious problems of nomenclature. A chabazite, for example, containing only Ca^{2+} ions in the exchange positions might be termed a Ca-chabazite, where "chabazite" designates the type of framework structure and "Ca" refers to the exchangeable cation population. Structurally, the mineral gmelinite is closely related to chabazite and for years was thought to be the sodium analogue of chabazite. The framework structures are significantly different from one another, however, and Na-chabazite is not the same as gmelinite or even Na-gmelinite. On the other hand, the mineral harmotome is actually a Ba-phillipsite in that it possesses basically the same crystal structure as phillipsite but contains a predominance of barium in the exchange positions. The name, however, is well established in the literature and probably should be retained.

Loosely bound water is also present in the structures of all natural zeolites and ranges from 10-20% of the dehydrated phase. Part or all of this water is given off continuously and reversibly on heating from room temperature to about 350°C. Once the water is removed, the cations fall back to positions on the inner surface of the channels and central cavities. The dehydration (or activation) of a zeolite is an endothermic process; conversely, rehydration is exothermic. Consequently, at a given temperature the water content is dependent on the P_{H_2O} of the atmosphere to which the zeolite is exposed. Outcrop specimens of zeolitic tuff in the Mohave Region of southern California, for example, are noticeably dehydrated and cause a slight burning sensation when placed on the tongue; a positive, but somewhat unpleasant field test.

Crystal Structure

Whereas the framework structures of quartz and feldspar are dense and tightly packed (S.G. = 2.6-2.7), those of zeolite minerals are remarkably open (S.G. = 2.0-2.2), and void volumes of dehydrated species as great as 50% are known (see Table 1). Although the chemical compositions are very similar, each zeolite species possesses its own unique crystal structure, and hence, its own set of physical and chemical properties. Most structures, however, can be visualized as SiO_4 and AlO_4 tetrahedra linked together in a simple geometrical form, such as that shown in Figure 2a. This particular polyhedron is known as a truncated cuboctahedron. It is more easily seen by considering only lines joining the midpoints of each tetrahedron, as shown in Figure 2b. Individual polyhedra may be connected in several ways; for example, by double

4

Table 1. Typical Formulae and Selected Physical Properties of Important Zeolites.

Zeolite	Typical Unit-Cell Formula[1]	Crystal System	Void Volume[1]	Specific Gravity[6]	Channel Dimensions[1]	Thermal Stability	Ion-Exchange Capacity[2]	Habit in Sedimentary Rocks
ANALCIME	$Na_{16}(Al_{16}Si_{32}O_{96})\cdot16H_2O$	Cubic	18%	2.24-2.29	2.6 Å	High	4.54 meq/g	trapezohedra
CHABAZITE	$(Na_2,Ca)_6(Al_{12}Si_{24}O_{72})\cdot40H_2O$	Hexagonal	47	2.05-2.10	3.7 x 4.2	High	3.81	rhombs
CLINOPTILOLITE	$(Na_4K_4)(Al_8Si_{40}O_{96})\cdot24H_2O$	Monoclinic	39?	2.16	3.9 x 5.4	High	2.54	laths and plates
ERIONITE	$(Na,Ca_{.5},K)_9(Al_9Si_{27}O_{72})\cdot27H_2O$	Hexagonal	35	2.02-2.08	3.6 x 5.2	High	3.12	hexagonal rods and bundles of rods
FAUJASITE[4]	$Na_{58}(Al_{58}Si_{134}O_{384})\cdot240H_2O$	Cubic	47	1.91-1.92	7.4	High	3.39	N.A.
FERRIERITE	$(Na_2Mg_2)(Al_6Si_{30}O_{72})\cdot18H_2O$	Orthorhombic	-	2.14-2.21	4.3 x 5.5 / 3.4 x 4.8	High	2.33	laths and rods
HEULANDITE	$Ca_4(Al_8Si_{28}O_{72})\cdot24H_2O$	Monoclinic	39	2.10-2.20	4.0 x 5.5 / 4.4 x 7.2 / 4.1 x 4.7	Low	2.91	laths
LAUMONTITE	$Ca_4(Al_8Si_{16}O_{48})\cdot16H_2O$	Monoclinic	34	2.20-2.30	4.6 x 6.3	Low	4.25	laths
MORDENITE	$Na_8(Al_8Si_{40}O_{96})\cdot24H_2O$	Orthorhombic	28	2.12-2.15	2.9 x 5.7	High	2.29	needles and fibers
NATROLITE	$Na_{16}(Al_{16}Si_{24}O_{80})\cdot16H_2O$	Orthorhombic	23	2.20-2.26	6.7 x 7.0 / 2.6 x 3.9	Low	5.26	-
PHILLIPSITE	$(Na,K)_{10}(Al_{10}Si_{22}O_{62})\cdot20H_2O$	Orthorhombic	31	2.15-2.20	4.2 x 4.4 / 2.8 x 4.8 / 3.3	Low	3.87	rods and laths
WAIRAKITE	$Ca_8Al_{16}Si_{32}O_{96}\cdot16H_2O$	Monoclinic	20?	2.26	-	High	4.61	-
LINDE A[3]	$Na_{12}(Al_{12}Si_{12}O_{48})\cdot27H_2O$	Cubic	47	1.99	4.2	High	5.48	N.A.
LINDE X[3]	$Na_{86}(Al_{86}Si_{106}O_{384})\cdot264H_2O$	Cubic	50	1.93	7.4	High	4.73	N.A.

[1]Taken mainly from Breck, 1974; Meier and Olson, 1971. Void volume is determined from water content.
[2]Calculated from unit-cell formula.
[3]Linde A and Linde X are synthetic phases.
[4]Faujasite is rare and not found in sedimentary rocks.
[5]N.A. = not applicable
[6]Taken mainly from Breck, 1974; Deer, Howie, and Zussman, 1963. Mostly determined on crystals from amygdales in basalt.

5

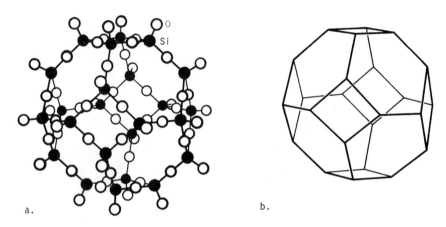

a.   b.

Figure 2. Simple polyhedron of silicate and aluminate tetrahedra. (a) Ball and peg
model of truncated cuboctahedron.   (b) Line drawing of truncated cuboctahedron; lines connect centers of tetrahedra.

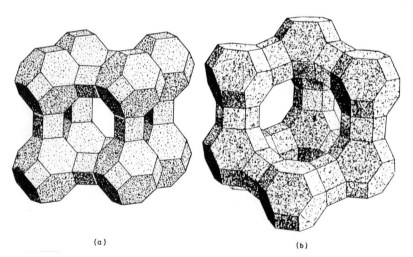

(a)   (b)

Figure 3. Arrangements of simple polyhedra to enclose large central cavities.
(a)   Truncated cuboctahedra connected by double 4-rings of oxygen in
structure of synthetic zeolite A.   (b) Truncated cuboctahedra connected
by double 6-rings of oxygens in structure of faujasite. (From Meier,
1968.)

6

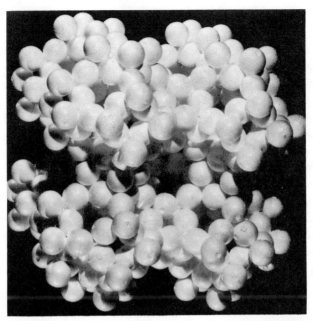

Figure 4a.   Solid-sphere model of crystal structure of synthetic zeolite A.   (From Mumpton and Fishman, 1977.)

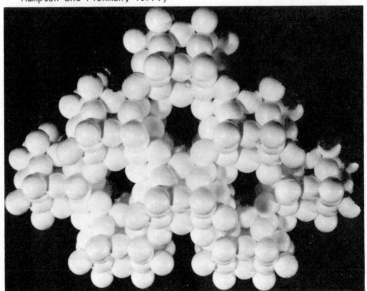

Figure 4b.   Solid-sphere model of crystal structure of chabazite.   (From Mumpton and Fishman, 1977.)

4-rings of oxygen ions, as shown in Figure 3a, or by double 6-rings as shown in Figure 3b, the framework structures of synthetic zeolite A and the mineral faujasite, respectively. Solid-sphere models of synthetic zeolite A and the mineral chabazite are illustrated in Figures 4a and 4b.

From an examination of Figures 2 to 4, it is evident that there is considerable void space within both the simple polyhedra building blocks and within the larger frameworks formed by several polyhedra. The passageways leading into the simple polyhedra are too small for all but the smallest molecules to pass; however, ports or channels up to 8 Å in diameter lead into the large, three-dimensional cavities, as shown in Figures 3 and 4, allowing species as large as benzene or cyclohexane to be adsorbed. The system of channels and cavities is different in each zeolite structure, as are the effective sizes of the entry ports, giving rise to a wide variety of materials each capable of screening molecules and cations by molecular and ion sieving in slightly different manners. (See Chapter 8.)

HISTORICAL

In the 200 years following their discovery, zeolite minerals were regarded by most of the geological profession as mineralogical curiosities. Despite their occurrence in almost every basalt flow, few investigations of major geological significance were carried out on them during this period, and zeolites were relegated for the most part to museum cases and collection drawers. Here they remained, collecting dust both in a real and in a scientific sense. Textbooks of the period (as well as most published today!) invariably described these minerals as having crystallized from late-stage magmatic solutions and offered little additional information. It was universally accepted that zeolites almost always occurred as vesicule fillings in basic extrusive rocks.

The chemists, however, were not so reticent in studying zeolites, as evidenced as early as 1857 by Damour's discovery that zeolites could be reversibly dehydrated without destruction of the crystal, and by Friedel's observation (1896) that various liquids, such as benzene, alcohol, chloroform, and mercury could be occluded by the dehydrated material. In 1909, Grandjean demonstrated the adsorption properties of zeolites using chabazite and such gases as hydrogen, air, ammonia, H_2S, and iodine, and in 1925, Weigel and Steinhoff noted that dehydrated zeolites would adsorb small organic molecules, but reject larger ones, a phenomenon described in 1932 by McBain as "molecular sieving." In the two decades that followed, dozens of papers on the dehydration, adsorption, and ion-exchange properties of zeolite minerals appeared in the chemical literature, mostly from the laboratories of R. M. Barrer in London and J. Sameshima in Japan. Much of this work was focused on the zeolites mordenite and chabazite which appeared to have the greatest adsorption capacities of the zeolites known at that time; however, the rarity of these zeolites, or of any zeolites for that matter, precluded the development of large-scale industrial processes based on natural materials. Zeolites were sufficiently abundant in amygdaloidal basalts for experimental purposes, but there seemed to be no method that could economically extract zeolites from such bodies to support a commercial process.

The non-existence of commercial deposits of natural zeolites caused the chemists to turn to synthesis as a means of obtaining a steady supply of zeolite materials, and between 1944 and 1960, major efforts went into the low-temperature hydrothermal synthesis of crystalline zeolites. Based on earlier experiments by Barrer, R. M. Milton of the Linde Division of Union Carbide Corporation began his own synthesis experiments in Tonawanda, New York. His initial runs, aimed at making chabazite because of its potential in the separation of oxygen and nitrogen from air, were unsuccessful. What was produced, however, was an entirely new type of zeolite, possessing adsorption and molecular sieve properties even better than those of chabazite. This new zeolite is the Linde type A zeolite, the mainstay of the synthetic molecular sieve business (Milton, 1959, 1968).

The 1950s found Union Carbide Corporation and other industrial laboratories deeply engaged in developing processes and markets for their growing molecular sieve business. It was during this period that the geological profession began to hear rumors that zeolites could be found in rocks other than basalt flows, and that in some places, zeolite minerals made up almost 100% of the rock! Hints of such non-basalt, non-igneous occurrences had found their way into the geological literature before this, but in general these reports were ignored by the profession. In 1914, Albert Johannsen found what he thought were fine-grained zeolites making up a large portion of Eocene tuff beds in the Uintah Basin of Utah, Colorado, and Wyoming. In 1928, W. H. Bradley and C. S. Ross both described zeolites in saline-lake deposits of Wyoming and Arizona, respectively, and in 1933, two U. S. Geological Survey geologists, M. N. Bramlette and E. Posjnak, reported several occurrences of clinoptilolite in vitric tuffs in various parts of the western United States. Kerr (1931) noted zeolites in California bentonites, and Fenner (1936) described extensive zones of what he identified as heulandite in Yellowstone drill cores. The deep-sea zeolites reported by Murray and Renard in 1891 in the report of the H.M.S. Challenger predate all of the above, but this work was also not known extensively until recently. Probably the earliest record of zeolites in sedimentary rocks was published in 1876 by Lew, who described chabazite from

9

a tuff bed near Bowie, Arizona, the site of current zeolite mining operations (see Sheppard et al., 1976).

In the 1950's, although the geological world was generally unaware of it, zeolites were being found in rocks other than igneous basalts, mostly in bedded tuffs in lacustrine and marine environments. Influenced by Coombs' discovery (1954) of several different types of zeolites in low-grade metamorphic rocks of New Zealand, Sudo's work in Japan (1950) on the zeolites of the Green Tuff, the several reports of "mordenite-like" zeolites in sedimentary rocks of the Soviet Union, and Sersale's early work (1958) on the zeolitic "tuffo giallo napoletano" in Italy, a few far-sighted geologists in the United States began to examine tuffaceous sediments and altered volcanic ashes by x-ray diffraction techniques. The results speak for themselves. What appeared in hand specimen to be ordinary, fine-grained, altered glasses, were shown to contain up to 95% of a single zeolite (e.g., Ames, Sand, and Goldich, 1958; Hay, 1962; Van Houten, 1964; Deffeyes, 1959a; Mumpton, 1960). Most of the zeolites in altered tuffs are extremely fine grained and defy reliable microscopic characterization. Particle sizes range from less than 0.1 µm to a few µm; however, as shown by scanning electron microscopy, crystal shapes are well developed and generally mimic those of megascopic crystals found in basalt amygdules (see Figures 5 to 12). As a result of such investigations, several million-ton size deposits of "rare" mineral erionite were discovered in 1957 in lacustrine deposits of central Nevada (Deffeyes, 1959b) and in southwest Oregon, and the zeolite clinoptilolite suddenly became one of the most common minerals in tuffaceous sedimentary rocks.

Thus, in the late 1950s, while the chemists were busy synthesizing zeolites and finding uses for them, the geologists were realizing that many zeolite minerals could be found in mineable quantities in nature. When each group learned of the other's activities, the zeolite story picked up its pace. The geologists were stimulated because not only were their discoveries of significance in deciphering the origin of volcanogenic sedimentary rocks, but the zeolite constitutents of these rocks were valuable from a commercial point of view as well. Simultaneously, the chemists were now able to design uses based on relatively inexpensive natural materials to complement the more costly synthetic products, thereby creating entirely new markets for zeolite materials.

The discovery that zeolite minerals were formed on a large scale by the reaction of volcanic tuffs and tuffaceous sedimentary rocks with lacustrine, marine, or ground waters in a host of geological environments was a milestone in the geological sciences; however, the realization that such materials were also capable of being utilized in numerous areas of industrial and agricultural technology provided the

Figure 5. Scanning electron micrograph of clinoptilolite from a lacustrine tuff near Castle Creek, Idaho. Note characteristic monoclinic symmetry of blades and laths, some of which are similar to the coffin shape of megascopic heulandite from basalt amygdules. (From Mumpton and Ormsby, 1976.)

Figure 6. Scanning electron micrograph of erionite needles from a saline-lake deposit near Eastgate, Nevada. Needles are 10 to 20 μm in length and about 1 μm thick. (From Mumpton and Ormsby, 1976.)

Figure 7. Scanning electron micrograph of erionite from a tuff near Hector, California. The bundles are about 20 μm long and about 10 μm wide and contain several hundred needles, each about 0.5 μm in diameter. (From Mumpton and Ormsby, 1976.)

Figure 8. Scanning electron micrograph of chabazite rhombs from a lacustrine tuff along Beaver Divide, Wyoming. The crystals are several μm in diameter and commonly intergrown. (From Mumpton and Ormsby, 1976.)

Figure 9. Scanning electron micrograph of phillipsite from deep-sea sediments. Prisms are about 10 to 15 μm long and about 2 to 3 μm wide. (Courtesy of R. E. Garrison.)

Figure 10. Scanning electron micrograph of analcime crystals displaying typical trapazohedral symmetry from Ischia, Italy. (From Mumpton and Ormsby, 1976.)

Figure 11. Scanning electron micrograph of mordenite needles from a lacustrine tuff near Lovelock, Nevada. Needles are about 5 to 20 μm in length and 0.5 to 1 μm wide. (From Mumpton and Ormsby, 1976.)

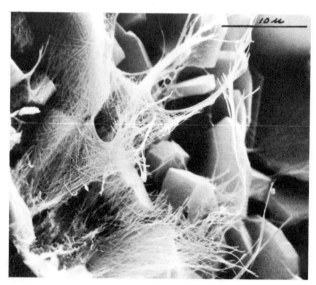

Figure 12. Scanning electron micrograph of intertwined mordenite fibers with clinoptilolite from an ash-flow tuff near Monolith, California. (From Mumpton and Ormsby, 1976.)

14

impetus for the exploration and development programs that have been carried out in recent years in many countries. In less than two decades, the status of the zeolite group of minerals has changed from that of museum curiosity to an important industrial mineral commodity. The commercial use of natural zeolites in still in its infancy; however, more than 300,000 tons of zeolitic tuff is currently mined each year in the United States, Japan, Italy, Bulgaria, Hungary, Germany, Korea, and Mexico. Natural zeolites are used as filler in the paper industry, as lightweight aggregate, in pozzolanic cements and concrete, as dietary supplements in animal husbandry, as ion exchangers in pollution-abatement processes and water purification, in the separation of oxygen and nitrogen from air, as reforming petroleum catalysts, in fertilizers and soil conditioners, and as acid-resistant adsorbents in gas drying and purification. In this era of energy and resource conservation and environmental concern, the attractive chemical and physical properties of natural zeolites will undoubtedly be harnessed even more in the future in the solution to these and other problems.

REFERENCES

Ames, L.L., Sand, L.B., and Goldich, S.S. (1958) A contribution on the Hector, California bentonite deposit: Econ. Geol. 53, 22-37.

Breck, D.W. (1974) Zeolite Molecular Sieves: John Wiley & Sons, Inc., New York, 771 pp.

Bradley, W.H. (1928) Zeolite beds in the Green River Formation: Science 67, 73-74.

Bramlette, M.N. and Posnjak, E. (1933) Zeolitic alteration of pyroclastics: Am. Mineral. 18, 167-181.

Brush, G.J. and Penfield, S.L. (1926) Manual of Determinative Mineralogy, 16th ed., rev., John Wiley & Sons, New York, p. 282.

Coombs, D.S. (1954) The nature and alteration of some Triassic sediments from Southland, New Zealand: Trans. Royal Soc. New Zealand 82, 65-109.

Cronstdet, A.F. (1756) On em obekant barg art, som kallas Zeolites: Akad. Handl. Stockholm 17, 120-123.

Damour, A. (1857) Recherches sur les propriétés hygroscopiques des mineraux de la famille des zéolites: C.R. Acad. Sci. 44, 975-980.

Deer, W.A., Howie, R.A., and Zussman, J. (1963) Rock-forming Minerals, Vol. 4, Framework Silicates: Longman, London, 351-428.

Deffeyes, K.G. (1959a) Zeolites in sedimentary rocks: J. Sed. Petrol. 29, 602-609.

Deffeyes, K.G. (1959b) Erionite from Cenozoic tuffaceous sediments: Am. Mineral. 44, 501-509.

Fenner, C.N. (1936) Bore-hole investigations in Yellowstone Park: J. Geol. 44, 225-315.

Friedel, G. (1896) Sur quelques porpietes nouvelles des zeolithes: Bull. Soc. franc. Mineral. Cristallogr. 19, 94-118.

Grandjean, F. (1909) Optical study of the adsorption of heavy vapors on certain zeolites: Comptes rend. 149, 866-868.

Hay, R.L. (1962) Origin and diagenetic alteration of the lower part of the John Day Formation near Mitchell, Oregon: In, Engel, A.E.J., James, H.L., and Leonard, B.F., Eds., Petrologic Studies: A Volume in Honor of A.F. Buddington, Geol. Soc. Am., 191-216.

Johannsen, A. (1914) Petrographic analysis of the Bridger, Washakie, and other Eocene Formations of the Rocky Mountains: Bull. Am. Mus. Nat. Hist. 33, 209-222.

Kerr, P.F. (1931) Bentonite from Ventura, California: Econ. Geol. 26, 153-168.

Loew, O. (1875) Report upon mineralogical, agricultural, and chemical conditions observed in portions of Colorado, New Mexico, and Arizona in 1873: Pt. 6, U.S. Geogr. and Geol. Explor. and Surv. West of 100th Meridian 3, 569-661.

McBain, J.W. (1932) The Sorption of Gases and Vapors by Solids: Routledge & Sons, London, Ch. 5.

Meier, W.M. (1968) Zeolite structures: Molecular Sieves, Soc. Chem. Ind., London, 10-27.

Meier, W.M. and Olson, D.H. (1971) Zeolite frameworks: Adv. in Chem. Ser. 101, Am. Chem. Soc., 155-170.

Milton, R.M. (1959) Molecular sieve adsorbents: U.S. Patent 2,882,243, April 14.

Milton, R.M. (1968) Commercial development of molecular sieve technology: In, Molecular Sieves, Soc. Chem. Ind., London, 199-203.

Mumpton, F.A. (1960) Clinoptilolite redefined: Am. Mineral. 45, 351-369.

Mumpton, F.A. and Ormsby, W.C. (1976) Morphology of zeolites in sedimentary rock by scanning electron microscopy: Clay and Clay Minerals 24, 1-23.

Mumpton, F.A. and Fishman, P.H. (1977) The application of natural zeolites in animal science and aquaculture: J. Anim. Sci. 45, 1188-1203.

Murray, J. and Renard, A.F. (1891) Deep sea deposits: Vol. 5, Report on the Scientific Results of the Voyage of the HMS Challenger during the years 1873-76, Eyre & Spottiswoode, London, 525 pp.

Northrup, M.A. (1941) Home laboratory tests for the identification of zeolites and related silicates: Rocks and Minerals 16, 275-279.

Ross, C.S. (1928) Sedimentary analcite: Am. Mineral. 26, 627-629.

Sersale, R. (1958) Genesi e constituzione del tufo giallo napoletano: Rend. Accad. Sci. Fische Mat. (Napoli) 25, 181-207.

Sheppard, R.A., Gude, A.J., 3d, and Edson, G.M. (1976) Bowie zeolite deposit, Cochise and Graham Counties, Arizona: In Natural Zeolites: Occurrence, Properties, Use, L.B. Sand and F.A. Mumpton, Eds., Pergamon Press, Elmsford, New York, 319-328.

Sudo, T. (1950) Mineralogical studies on the zeolite-bearing pumice tuffs near Yokotemachi, Akita Prefecture: J. Geol. Soc. Japan 56, 13-16.

Van Houten, F.B. (1964) Tertiary geology of the Beaver Rim area, Fremont and Natrona
     Counties, Wyoming: U. S. Geol. Surv. Bull. 1164, 99 pp.

Weigel, O. and Steinhof, E. (1925) Adsorption of organic liquid vapors by chabazite:
     Z. Kristallogr. 61, 125-154.

Chapter 2

CRYSTAL STRUCTURE AND CHEMISTRY OF NATURAL ZEOLITES

Edith M. Flanigen

INTRODUCTION

Zeolites are crystalline, hydrated aluminosilicates of alkali and alkaline earth
metal elements; in particular, of sodium, potassium, magnesium, calcium, strontium, and
barium. Structurally they are framework aluminosilicates consisting of infinitely
extending three-dimensional networks of AlO_4 and SiO_4 tetrahedra linked to each other
by the sharing of all oxygens. Zeolites are most closely related to the feldspar and
feldspathoid groups of silicate minerals, and all three groups illustrate the substitution of a monovalent cation plus aluminum for silicon in the basic formula of the
silica minerals, as follows:

$$(SiO_2)_n \xrightarrow{\quad xM + xAl \quad} M_x(AlO_2)_x(SiO_2)_{n-x}$$

Rearranging the empirical formula listed on page 3, a crystallographic unit-cell
formula can be developed as shown below:

$$M_{a/n}((AlO_2)_a(SiO_2)_b) \cdot w\ H_2O$$

where M is a cation of valence n, w is the number of water molecules, and a and b are
small whole numbers. The sum (a + b) is the total number of tetrahedra in the unit cell,
and the ratio b/a varies from 1 to 5. Thus, the zeolite structure may be divided into
three components: the aluminosilicate framework, interconnected void spaces in the
framework containing metal cations, M, and water molecules which are present as an
occluded phase.

Zeolites comprise the largest group of tektosilicates; more than 35 different
framework topologies are known, and an infinite number are possible. Nearly 40 species
of zeolite minerals have been described, and about 100 different types of synthetic
zeolites have been made in the laboratory.

CRYSTAL STRUCTURES

It is convenient to describe and classify zeolite structures in terms of fundamental building units, as listed in Table 1. These units include the primary building
unit of TO_4 tetrahedra, so-called secondary building units (SBU) which consist of both
single rings of 4, 5, 6, 8, 10, and 12 tetrahedra and double rings of 4, 6, and 8
tetrahedra, and larger symmetrical polyhedra described in terms of Archimedean solids.
Several of the secondary building units and the larger symmetrical polyhedra are shown
in Figure 1. Methods used to depict structural units in zeolites and for building
models to illustrate crystal structures of zeolites are summarized in Appendix 1.

Detailed structural classifications of natural and synthetic zeolites have been
proposed by Smith (1963), Fischer and Meier (1965), and Breck (1974). Although Meier
(1968) was the first to propose a classification based on secondary building units
contained in the structure, Breck's classification is based on a combination of frame-

Table 1. Building Units in Zeolite Structures[1]

Primary Building Unit - Tetrahedron (TO_4)

Tetrahedron of four oxygen ions with a
central ion (T) of Si^{+4} or Al^{+3}

Secondary Building Units (SBU)

Rings: S-4, S-5, S-6, S-8, S-10, S-12

Double Rings: D-4, D-6, D-8

Larger Symmetrical Polyhedra

Truncated Octahedron (T.O.) or Sodalite Unit

11-Hedron or Cancrinite Unit

14-Hedron II or Gmelinite Unit

[1]From Flanigen et al., 1971.

20

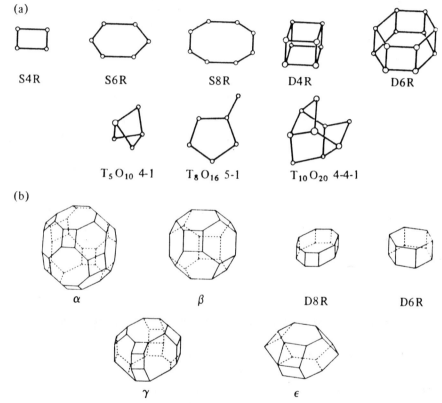

Figure 1. (a) The secondary building units (SBU) in zeolite structures according to Meier (1968). Only the positions of tetrahedral (T) silicons and aluminums are shown. Oxygen atoms lie near the connecting solid lines, which are not intended to mean bonds. The 4-1 unit is based on the configuration of 5 tetrahedra present in the structures of Group 5. The 5-1 unit is based on the configuration of 5-rings found in Group 6. The 4-4-1 unit is based on the configuration of tetrahedra found in Group 7. (b) Some polyhedra in zeolite frameworks: α (26-hedron Type I) or truncated cuboctahedron; β (14-hedron Type I) or truncated octahedron; δ or double 8-ring; D6R or double 6-ring (hexagonal prism); γ or 18-hedron; and ϵ or the 11-hedron. (From Breck, 1974, Fig. 2.18.)

work topology as well as secondary building units and will be used in this chapter.
Breck's division of zeolites into seven groups based on these criteria is shown below:

Group	Secondary Building Unit
1	Single 4-ring (S4R)
2	Single 6-ring (S6R)
3	Double 4-ring (D4R)
4	Double 6-ring (D6R)
5	Complex 4-1, T_5O_{10} unit
6	Complex 5-1, T_8O_{16} unit
7	Complex 4-4-1, $T_{10}O_{20}$ unit

His detailed compilation of zeolite structures into these groups is shown in Table 2.
Meier's original groups are listed in Table 3 which includes additional crystallographic data not found in Table 2. Meier's table showing the relationship of secondary
building units present in various zeolite structures is reproduced as Table 4.

Group 1 (S4R)

Group 1 consists of framework structures which are formed from single 4-rings.
It includes the important natural zeolites analcime, phillipsite, gismondine, and laumontite. The structures of paulingite, harmotome, and yugawaralite also fall into
this group.

Analcime. A stereogram of the framework topology of analcime is shown in Figure 2
and is described in the figure legend. Analcime is a common zeolite in both igneous
basalts and in sedimentary rocks and has been synthesized on numerous occasions. Although it is relatively complex, the crystal structure of analcime was one of the
first to have been determined. The minerals analcime, wairakite, viseite, and kehoeite
have basically the same framework structure. The name analcime is used to designate
the sodium-rich member of the analcime-wairakite series; wairakite--the calcium-rich
member. Whereas analcime is generally cubic, wairakite is only pseudo-cubic, crystal-
lizing in the monoclinic system. The water content of analcime varies linearly with
the silica content. As the silica content increases, the number of exchange cations
decreases. A summary of structural data for analcime and wairakite is shown in
Appendix II, Tables 1 and 2.

Viseite and kehoeite are varieties of analcime with disordered structures. They
are the only known zeolite minerals which contain ions other than silicon or aluminum

Table 2. Breck's Classification of Zeolites (Breck, 1974)

Name	Typical Unit Cell Contents	Type of Polyhedral Cage [a]	Framework Density, g/cc [b]	Void Fraction [c]	Type of Channels [d]	Free Aperture of Main Channels, Å [e]
Group 1 (S4R)						
Analcime	$Na_{16}[(AlO_2)_{16}(SiO_2)_{32}]\cdot 16\,H_2O$		1.85	0.18	One	2.6
Harmotome	$Ba_2[(AlO_2)_4(SiO_2)_{12}]\cdot 12\,H_2O$		1.59	0.31	Three	4.2 x 4.4
Phillipsite	$(K, Na)_{10}[(AlO_2)_{10}(SiO_2)_{22}]\cdot 20\,H_2O$		1.58	0.31	Three	4.2 x 4.4, 2.8 x 4.8
Gismondine	$Ca_4[(AlO_2)_8(SiO_2)_8]\cdot 16\,H_2O$		1.52	0.46	Three	3.1 x 4.4
P	$Na_6[(AlO_2)_6(SiO_2)_{10}]\cdot 15\,H_2O$		1.57	0.41	Three	3.5
Paulingite	$(K_2,Na,Ca,Ba)_{76}[(AlO_2)_{152}(SiO_2)_{520}]\cdot 700\,H_2O$	$\alpha,\gamma,\delta,$(10-hedron)	1.54	0.49	Three	3.9
Laumontite	$Ca_4[(AlO_2)_8(SiO_2)_{16}]\cdot 16\,H_2O$		1.77	0.34	One	4.6 x 6.3
Yugawaralite	$Ca_2[(AlO_2)_4(SiO_2)_{12}]\cdot 8\,H_2O$		1.81	0.27	Two	3.6 x 2.8
Group 2 (S6R)						
Erionite[f]	$(Ca,Mg,K_2,Na_2)_{4.5}[(AlO_2)_9(SiO_2)_{27}]\cdot 27\,H_2O$	ϵ, 23-hedron	1.51	0.35	Three	3.6 x 5.2
Offretite[f]	$(K_2,Ca)_{2.7}[(AlO_2)_{5.4}(SiO_2)_{12.6}]\cdot 15\,H_2O$	ϵ, 14-hedron (II)	1.55	0.40	Three	3.6 x 5.2, ‖a; 6.4, ‖c
T	$(Na_{1.2}K_{2.8}[(AlO_2)_4(SiO_2)_{14}]\cdot 14\,H_2O$	ϵ, 23, 14-hedron	1.50	0.40	Three	3.6 x 4.8
Levynite[m]	$Ca_3[(AlO_2)_6(SiO_2)_{12}]\cdot 18\,H_2O$	Ellipsoidal 17-hedron	1.54	0.40	Two	3.2 x 5.1
Omega[g]	$Na_{6.8}, TMA_{1.6}[(AlO_2)_8(SiO_2)_{28}]\cdot 21\,H_2O$	14-hedron (II)	1.65	0.38	One	7.5
Sodalite Hydrate	$Na_6[(AlO_2)_6(SiO_2)_6]\cdot 7.5\,H_2O$	β	1.72	0.35	Three	2.2
Losod	$Na_{12}[(AlO_2)_{12}(SiO_2)_{12}]\cdot 19\,H_2O$	ϵ, 17-hedron	1.58	0.33	Three	2.2
Group 3 (D4R)[n]						
A	$Na_{12}[(AlO_2)_{12}(SiO_2)_{12}]\cdot 27\,H_2O$	α, β	1.27	0.47	Three	4.2
N-A	$Na_4TMA_3[(AlO_2)_7(SiO_2)_{17}]\cdot 21\,H_2O$	α, β	1.3	0.5	Three	4.2
ZK-4	$Na_8TMA[(AlO_2)_9(SiO_2)_{15}]\cdot 28\,H_2O$	α, β	1.3	0.47	Three	4.2

Table 2 (continued)

Name	Typical Unit Cell Contents	Type of Polyhedral Cage[a]	Framework Density, g/cc[b]	Void Fraction[c]	Type of Channels[d]	Free Aperture of Main Channels, A[e]
Group 4 (D6R)						
Faujasite	$(Na_2, K_2, Ca, Mg)_{29.5}[(AlO_2)_{59}(SiO_2)_{133}] \cdot 235\ H_2O$	β, 26-hedron(II)	1.27	0.47	Three	7.4
X	$Na_{86}[(AlO_2)_{86}(SiO_2)_{106}] \cdot 264\ H_2O$	β, 26-hedron (II)	1.31	0.50	Three	7.4
Y	$Na_{56}[(AlO_2)_{56}(SiO_2)_{136}] \cdot 250\ H_2O$	β, 26-hedron (II)	1.25 - 1.29	0.48	Three	7.4
Chabazite	$Ca_2[(AlO_2)_4(SiO_2)_8] \cdot 13\ H_2O$	20-hedron	1.45	0.47	Three	3.7 x 4.2
Gmelinite	$Na_8[(AlO_2)_8(SiO_2)_{16}] \cdot 24\ H_2O$	14-hedron (II)	1.46	0.44	Three	3.6 x 3.9, ‖a; 7.0, ‖c
ZK-5[o]	$(R, Na_2)_{15}[(AlO_2)_{30}(SiO_2)_{66}] \cdot 98\ H_2O$	α, γ	1.46	0.44	Three	3.9
L[h]	$K_9[(AlO_2)_9(SiO_2)_{27}] \cdot 22\ H_2O$	ε	1.61	0.32	One	7.1
Group 5 (T_5O_{10})[i]						
Natrolite	$Na_{16}[(AlO_2)_{16}(SiO_2)_{24}] \cdot 16\ H_2O$		1.76	0.23	Two	2.6 x 3.9
Scolecite	$Ca_8[(AlO_2)_{16}(SiO_2)_{24}] \cdot 24\ H_2O$		1.75	0.31	Two	2.6 x 3.9
Mesolite	$Na_{16}Ca_{16}[(AlO_2)_{48}(SiO_2)_{72}] \cdot 64\ H_2O$		1.75	0.30	Two	2.6 x 3.9
Thomsonite	$Na_4Ca_8[(AlO_2)_{20}(SiO_2)_{20}] \cdot 24\ H_2O$		1.76	0.32	Two	2.6 x 3.9
Gonnardite	$Na_4Ca_2[(AlO_2)_8(SiO_2)_{12}] \cdot 14\ H_2O$		1.74	0.31	Two	2.6 x 3.9
Edingtonite	$Ba_2[(AlO_2)_4(SiO_2)_6] \cdot 8\ H_2O$		1.68	0.36	Two	3.5 x 3.9
Group 6 (T_8O_{16})[j]						
Mordenite	$Na_8[(AlO_2)_8(SiO_2)_{40}] \cdot 24\ H_2O$		1.70	0.28	Two	6.7 x 7.0, ‖c; 2.9 x 5.7, ‖b
Dachiardite	$Na_5[(AlO_2)_5(SiO_2)_{19}] \cdot 12\ H_2O$		1.72	0.32	Two	3.7 x 6.7, ‖b; 3.6 x 4.8, ‖c
Ferrierite	$Na_{1.5}Mg_2[(AlO_2)_{5.5}(SiO_2)_{30.5}] \cdot 18\ H_2O$		1.76	0.28	Two	4.3 x 5.5 ‖c; 3.4 x 4.8 ‖b
Epistilbite	$Ca_3[(AlO_2)_6(SiO_2)_{18}] \cdot 18\ H_2O$		1.76	0.25	Two	3.2 x 5.3, ‖a; 3.7 x 4.4, ‖c
Bikitaite	$Li_2[(AlO_2)_2(SiO_2)_4] \cdot 2\ H_2O$		2.02	0.23	One	3.2 x 4.9

Table 2 (continued)

Name	Typical Unit Cell Contents	Type of Polyhedral Cage [a]	Framework Density, g/cc [b]	Void Fraction [c]	Type of Channels [d]	Free Aperture of Main Channels, A [e]
Group 7 ($T_{10}O_{20}$) [k]						
Heulandite	$Ca_4[(AlO_2)_8(SiO_2)_{28}] \cdot 24\ H_2O$		1.69	0.39	Two	4.0 x 5.5, $\|$a 4.0 x 7.2, $\|$c
Clinoptilolite	$Na_6[(AlO_2)_6(SiO_2)_{30}] \cdot 24\ H_2O$		1.71	0.34	?	?
Stilbite	$Ca_4[(AlO_2)_8(SiO_2)_{28}] \cdot 28\ H_2O$		1.64	0.39	Two	4.1 x 6.2, $\|$a 2.7 x 5.7, $\|$c
Brewsterite	$(Sr,\ Ba,\ Ca)_2[(AlO_2)_4(SiO_2)_{12}] \cdot 10\ H_2O$		1.77	0.26	Two	2.7 x 4.1 $\|$c 2.3 x 5.0 $\|$a

[a]Of the five space-filling solids of Federov, three (cube, hexagonal prism, and truncated octahedron) are found as polyhedral units in zeolite frameworks. The cube is the double 4-ring (D4R) as shown here. The double 6-ring (D6R) is the hexagonal prism or 8-hedron. The α-cage is the Archimedean semiregular, solid, truncated cuboctahedron referred to also as a 26-hedron, type I. The β-cage is the truncated octahedron or 14-hedron, type I. The γ-cage is the 18-hedron and the ϵ-cage the 11-hedron. Other polyhedral units are as given by Barrer (58).

[b]The framework density is based on the dimensions of the unit cell of the hydrated zeolite and framework contents only. Multiplication by 10 gives the density in units of tetrahedra/1000 A^3.

[c]The void fraction is determined from the water content of the hydrated zeolite.

[d]Refers to the network of channels which permeate the structure of the hydrated zeolite. Considerable distortion may occur in the group 5 and 7 zeolites upon dehydration.

[e]Based upon the structure of the hydrated zeolite.

[f]Erionite and offretite may also be considered to consist of double 6-rings linked by single 6-rings.

[g]Zeolite Ω may be considered to consist of single 6-rings linked by double 12-rings.

[h]Zeolite L consists of double 6-rings linked by single 12-rings.

[i]The T_5O_{10} refers to the unit of 5 tetrahedra as given by Meier for the 4-1 type of SBU. Fig. 67a (32).

[j]The T_8O_{16} unit refers to the characteristic configuration of tetrahedra shown in Fig. 72 (32).

[k]The $T_{10}O_{20}$ unit is the characteristic configuration of tetrahedra shown in Fig. 80 (32).

[l]Synthetic zeolites classified in group 1 include ZK-19 (35), W (36), and P-W (37) related to phillipsite, and various synthetic phases related to analcite—See Chapter 4.

[m]The synthetic zeolite, ZK-20 (38), is reported to have the levynite-type structure.

[n]Other zeolites with the A-type structure include zeolite α (39), ZK-21 (40) and ZK-22 (40).

[o]R = [1,4-dimethyl- 1,4-diazoniabicyclo (2,2,2) octane]$^{2+}$.

25

Table 3. Classification and Crystallographic Data of Zeolites (Meier, 1968)

Species	Idealised unit cell Contents	Crystal data (with unit cell constants in Å)	Isostructural species
Analcime group			
Analcime	$Na_{16}[Al_{16}Si_{32}O_{96}],16H_2O$	cubic Ia3d — $a=13\cdot7$	Wairakite, leucite, pollucite, viséite, kehoite
Natrolite group			
Natrolite	$Na_{16}[Al_{16}Si_{24}O_{80}],16H_2O$	orthorhomb. Fdd2 — $a=18\cdot30$ $b=18\cdot63$ $c=6\cdot60$	Mesolite, scolecite
Thomsonite	$Na_4Ca_8[Al_{20}Si_{20}O_{80}],24H_2O$	orthorhomb. Pnn2 — $a=13\cdot07$ $b=13\cdot08$ $c=13\cdot18$	Gonnardite
Edingtonite	$Ba_2[Al_4Si_6O_{20}],8H_2O$	orthorhomb. $P2_12_12$ (pseudotetrag. $P\bar{4}2_1m$) — $a=9\cdot54$ $b=9\cdot65$ $c=6\cdot50$	
Chabazite Group			
Gmelinite	$Na_8[Al_8Si_{16}O_{48}],24H_2O$	hexagonal $P6_3/mmc$ — $a=13\cdot75$ $c=10\cdot05$	
Chabazite	$Ca_2[Al_4Si_8O_{24}],13H_2O$	trigonal $R\bar{3}m$ — $a=9\cdot44$ $\alpha=94°28'$ (hexagonal) $a=13\cdot78$ $c=15\cdot06)$	
Erionite	(Ca etc.)$_{4\cdot5}$$[Al_9Si_{27}O_{72}],27H_2O$	hexagonal $P6_3/mmc$ — $a=13\cdot26$ $c=15\cdot12$	Linde T
Levynite	$Ca_3[Al_6Si_{12}O_{36}],18H_2O$	trigonal $R\bar{3}m$ — $a=10\cdot75$ $c=76°25'$ (hexagonal) $a=13\cdot32$ $c=22\cdot51)$	
Cancrinite hydrate	$Na_6[Al_6Si_6O_{24}],5H_2O$	hexagonal $P6_32$ (?) — $a=12\cdot7$ $c=5\cdot15$	
Sodalite hydrate	$Na_6[Al_6Si_6O_{24}],4H_2O$	cubic P43n — $a=8\cdot88$	Nosean, Zhdanov's G, danalite, tugtupite, (synthetic Ca-alumino-sulphate)
Phillipsite group			
Phillipsite	$(K,Na)_{10}[Al_{10}Si_{22}O_{64}],20H_2O$	orthorhombic B2mb — $a=9\cdot96$ $b=14\cdot25$ $c=14\cdot25$	
Gismondite	$Ca_4[Al_8Si_8O_{32}],16H_2O$	monoclinic $P2_1/c$ — $a=10\cdot02$ $b=10\cdot62$ $\beta=92°24'$ $c=9\cdot84$	Harmotome
Barrer's P1	$Na_6[Al_6Si_{10}O_{32}],15H_2O$	cubic Im3m — $a=10\cdot0$	Linde B, garronite (?)
Heulandite group			
Brewsterite	$(Sr,Ba,Ca)_2[Al_4Si_{12}O_{32}],10H_2O$	monoclinic $P2_1/m$ — $a=6\cdot77$ $b=17\cdot51$ $\beta=94°18'$ $c=7\cdot74$	
Heulandite	$Ca_4[Al_8Si_{28}O_{72}],24H_2O$	monoclinic Cm — $a=17\cdot71$ $b=17\cdot84$ $\beta=116°20'$ $c=7\cdot46$	Clinoptilolite
Stilbite	$Na_2Ca_4[Al_{10}Si_{26}O_{72}],28H_2O$	monoclinic C2/m — $a=13\cdot64$ $b=18\cdot24$ $\beta=128°0'$ $c=11\cdot27$	
Mordenite group			
Mordenite	$Na_8[Al_8Si_{40}O_{96}],24H_2O$	orthorhomb. Cmcm — $a=18\cdot13$ $b=20\cdot49$ $c=7\cdot52$	
Dachiardite	$Na_5[Al_5Si_{19}O_{48}],12H_2O$	monoclinic C2/m — $a=18\cdot73$ $b=7\cdot54$ $\beta=107°54'$ $c=10\cdot30$	
Epistilbite	$Ca_3[Al_6Si_{18}O_{48}],16H_2O$	monoclinic C2/m — $a=8\cdot92$ $b=17\cdot73$ $\beta=124°20'$ $c=10\cdot21$	
Ferrierite	$Na_2Mg_2[Al_6Si_{30}O_{72}],18H_2O$	orthorhomb. Immm — $a=19\cdot16$ $b=14\cdot13$ $c=7\cdot49$	
Bikitaite	$Li_2[Al_2Si_4O_{12}],2H_2O$	monoclinic $P2_1$ — $a=8\cdot61$ $b=4\cdot96$ $\beta=114°26'$ $c=7\cdot61$	
Faujasite group			
Faujasite	$(Na_2,Ca)_{29}[Al_{58}Si_{134}O_{384}],256H_2O$	cubic Fd3m — $a=24\cdot7$	Linde X, Linde Y
Linde A	$Na_{96}[Al_{96}Si_{96}O_{384}],216H_2O$	cubic Fm3 (pseudo cell: cubic Pm3m $a=12\cdot32$) — $a=24\cdot64$	ZK-4
ZK-5	$Na_{30}[Al_{30}Si_{66}O_{192}],90H_2O$	cubic Im3m — $a=18\cdot7$	
Paulingite	$(K,Ca,Na)_{152}[Al_{152}Si_{520}O_{672}],\sim700H_2O$	cubic Im3m — $a=35\cdot1$	

Table 4. Secondary Building Units of Framework Zeolites (Meier, 1968)

Species	Single rings			Double rings		Complex or multiple units		
	4	6	8	4-4	6-6	4-1	5-1	4-4-1
Analcime	+	+						
Natrolite						+		
Thomsonite						+		
Edingtonite						+		
Gmelinite	+	+			+			
Chabazite	+	+	+		+			
Erionite	(+)	+	(+)					
Levynite	+	+						
Cancrinite	+	+						
Sodalite	+	+						
Phillipsite	+		+					
Gismondite	+		+	+				
Barrer P1	+		+					
Brewsterite	+							+
Heulandite								+
Stilbite								
Mordenite							+	
Dachiardite							+	
Epistilbite							+	
Ferrierite							+	
Bikitaite							+	
Faujasite		+	+	+	+			
Linde A		+	+		+			
ZK-5		+	+					
Paulingite			+					

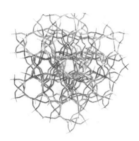 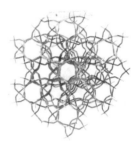

Figure 2.   Steroscopic view of the framework structure of analcime viewed in the direction of the 6-rings which form the nonintersecting channels parallel to
[111]. The simplest structural units in the framework are 4-rings and 6-
rings which are linked to create additional 8- and 12-rings. The 6-rings
are situated parallel to the [111] direction or 3-fold axis, and the 4-rings
are perpendicular to the 4-fold axes. The 8-rings, highly distorted, are
parallel to the (100) plane. The framework encompasses 16 cavities which
form continuous channels that run parallel to the 3-fold axes without in-
tersection. These channels are occupied by the water molecules. Twenty-
four smaller cavities are situated adjacent to these channels. (From
Breck, 1974, Fig. 2.4.)

in the tetrahedral framework, in both cases, phosphorus. Their chemical formulae are
$Na_2Ca_{10}((AlO_2)_{20}(SiO_2)_6(PO_2)_{10}(H_3O_2)_{12})\cdot 16H_2O$ and $Zn_{5.5}Ca_{2.5}((AlO_2)_{16})PO_2)_{16}(H_3O_2)_{16})\cdot$
$32H_2O$, respectively.

Phillipsite. Phillipsite is a principal member of a group of zeolites whose
structures are based on parallel 4-rings and 8-rings. These structures are closely
related to feldspar structures and have been classified by Smith (1963) on the basis
of the different types of 4-ring chains present in the structure. The framework of
linked 4-rings and 8-rings has an interconnected two-dimensional channel system.
Harmotome has the same aluminosilicate framework as phillipsite, but a different
chemical composition, with barium substituting for calcium and sodium in the exchange-
cation position. A stereogram of the framework structure of phillipsite and harmotome
is shown in Figure 3. Salient features of the structure are given in the figure legend.
Crystal structure data for these minerals are listed in Appendix II, Tables 3 and 4.

Gismondine. The framework structure of gismondine is also based on the four-
membered ring chain configuration of tetrahedra and also contains linked 4-rings and
8-rings. A stereogram of its framework topology is shown in Figure 4, and crystal
structure data are listed in Appendix II, Table 5. Figure 5 shows a comparison of

28

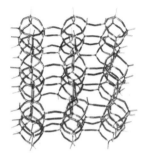 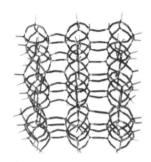

Figure 3. Stereoscopic view of the framework structure of phillipsite and harmotome. The view is parallel to the main 8-ring channel. The double chains are folded in an S-shaped configuration in the direction of the b-axis. The 4- and 8-rings are evident as well as the channels which parallel the a- and b-axes. The wider channel, parallel to a, is bounded by 8-rings with free dimensions of 4.2 x 4.8 Å and intersects channels parallel to the b-axis that have free dimensions of 2.8 and 4.8 Å. The channel intersections provide cavities. (From Breck, 1974, Fig. 2.5.)

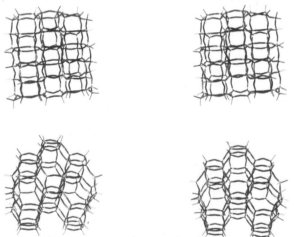

Figure 4. Stereoscopic view of gismondine. (a) View in the c direction parallel to the 4-fold axis. (b) View in the direction of the main 8-ring channel. Similar to harmontome and phillipsite, the channels in gismondine run parallel to the a and b directions. The main apertures in the channels are formed by 8-rings which have a free aperture size of 3-4 Å. In the channels each calcium ion is surrounded by 4 water molecules and coordinated with 2 of the framework oxygens. The water positions are only partially occupied, so that the average coordination number with water is four, to give a total average coordination of six. (From Breck, 1974, Fig. 2.6.)

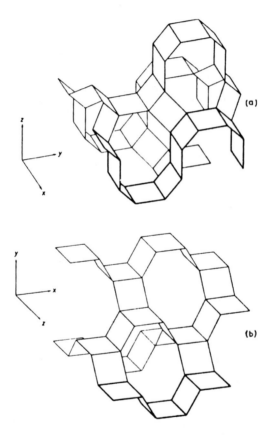

Figure 5. Skeletal diagrams of the frameworks of phillipsite or harmotome (a) and gismondine (b). (From Meier, 1968, Fig. 7.)

the different arrangements of four-membered ring chains in phillipsite, harmotome, and gismondine.

Laumontite. The framework of laumontite contains 4-, 6-, and 10-rings. The dehydration of laumontite occurs in stepwise fashion, with the partially dehydrated form being referred to as leonhardite. As seen in the stereogram of Figure 6, the framework structure of laumontite resembles that of analcime in that it consists of 4-rings and 6-rings. Crystal structure data for laumontite are listed in Appendix II, Table 6.

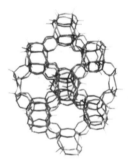 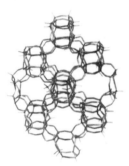

"Molecular Sieve Zeolites - I"

Figure 6. Framework of laumontite viewed parallel to a and showing the main channels formed by distorted 10-rings. (From Breck, 1974, Fig. 2.10.)

Group 2 (S6R)

Group 2 zeolites includes those structures having a single 6-ring as a common secondary building unit; hexagonal $(Al,Si)_6O_{12}$ (D-6) rings are also found in some of the frameworks. These structures can also be described according to the spatial arrangement of parallel 6-rings, linked through tilted 4-rings. The centers of the 6-rings are arranged like close-packed spheres in simple structures, using the sequences ABCABCABC..., ABABAB..., etc., into a large number of known and hypothetical structures, as illustrated in Figure 7. Members of this group include erionite, offretite, levynite, and the related feldspathoid, sodalite.

Erionite. Erionite has a hexagonal structure consisting of parallel D-6 rings. The framework is shown in Figure 8 and a space-filling model showing oxygen ions, in Figure 9. The erionite framework can also be considered in terms of the cancrinite

31

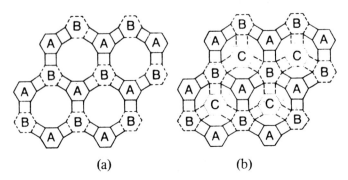

(a)   (b)

Figure 7. The AABB sequence (a) is characteristic of gmelinite, and the AABBCC (b)
arrangement is characteristic of chabazite. (From Breck, 1974, Fig. 2.22.)

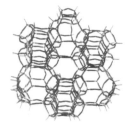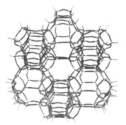

Figure 8. Stereoscopic view of the framework of erionite. The view is parallel to the
c-axis and shows the stacking sequence of the 6-rings. The sequence is AAB,
AAC. The framework consists of double 6-rings, D6R units, and S6R units
which are arranged in parallel planes perpendicular to the hexagonal axis.
Because of the stacking sequence, the c-dimension is 15 Å, i.e., 6 times
the 2.5 Å spacing of a single 6-ring.

The main cavities have internal diameters of 15.1 Å by 6.3 Å. Each of the
cavities has a single 6-ring at the top and bottom, which it shares with like
cavities above and below. The aperture between the cavities in the c-direction has a free diameter of 2.5 Å and is too small to permit diffusion of
most molecules. However, six 8-rings form apertures into any single cavity,
and these have free dimensions of 3.6 by 5.2 Å. Three are arranged in the
upper half of the cavity and three below. Each window or aperture formed
by these 8-rings is common to two cavities.

The ε- or cancrinite-type cages are linked in the c-direction by D6R units
(hexagonal prisms) in the configuration ε-D6R-ε-D64.... These columns are
crosslinked by single 6-rings perpendicular to c. In erionite the cages are
not symmetrically placed across the D6R units. (From Breck, 1974, Fig. 2.12.)

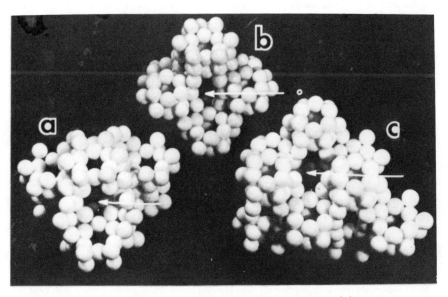

Figure 9.   Space-filling models of (a) erionite, (b) chabazite, (c) gmelinite.   The
8-ring aperture entering the channels is shown by the arrow.   (From Breck,
1974, Fig. 2.31.)

or ε-cages linked by D6R units in the c-direction. In adsorption and molecular sieve applications where the water has been removed from the zeolite structure, molecules must pass through 8-ring apertures to diffuse from one cavity to another. Continuous diffusion paths are available in the erionite structure for molecules of appropriate size.

The principal cations in natural erionite are potassium and calcium. In sedimentary varieties, the potassium ions show considerable resistance to ion exchange, indicating that they are locked within the structure, probably in the cancrinite or ε-cages. Crystal structure data are listed in Appendix II, Table 7.

Offretite. The offretite structure is closely related to that of erionite. A stereo-diagram of the framework topology of offretite is shown in Figure 10, and the crystal structure data are listed in Appendix II, Table 8. The structure consists of D-6 rings, gmelinite units, and cancrinite units. The framework of offretite contains an AABAAB... sequence of 6-rings, compared to AABAAC... for erionite. Some of the potassium ions in natural offretite also appear to be located within cancrinite cages.

Sodalite. Although the mineral sodalite is a feldspathoid rather than a zeolite and generally contains substantial quantities of chlorine, it is included in this discussion because of its structural relationship to the zeolites and because a hydrated variety is commonly encountered during zeolite synthesis studies. A sterogram of the sodalite structure is shown in Figure 11. The framework consists of a close-packed cubic array of polyhedral sodalite units (see also Figure 1b). Every 4- and 6-ring is shared by an adjacent sodalite unit. Crystal structure data for synthetic hydrated sodalite are given in Appendix II, Table 9.

Group 3 (D4R)

The secondary building unit common to framework structures of zeolites of Group 3 is a double 4-ring (D4R). Synthetic zeolite A is the only known structure type which is based on this unit.

Zeolite A. Zeolite A is one of the most important zeolites of the commercial molecular sieve business. It has been synthesized under a wide variety of conditions and from many different starting materials, but it has never been found in nature. The aluminosilicate framework of zeolite A contains two types of polyhedra, the double 4-rings of Figure 12 and the truncated octahedron or sodalite unit shown in Figure 13. The aluminosilicate framework is generated by placing the cubic D4R rings

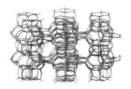 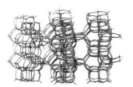

Figure 10. Stereoscopic view of the offretite framework looking perpendicular to c
along a. This view shows the stacking of D6Rs and ε-cages in the c-direction along with the gmelinite, 14-hedron cages. The main c-axis channels
are not shown in this view but resemble the channels in gmelinite and are
6.4 Å in diameter. (From Breck, 1974, Fig. 2.13.)

Figure 11. Stereodiagram of the framework topology of sodalite. (From Meier and Olson,
1971).

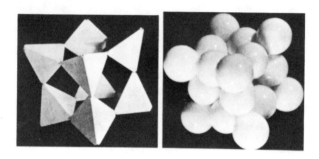

Figure 12.   The cubit unit of 8 tetrahedra as found in the framework structures of the Group 3 zeolites.   The photo on the left shows a model based on solid tetra-hedra and on the right, the space-filling model showing the positions of individual oxygen atoms.   This is referred to as the double 4-ring or D4R unit.   (From Breck, 1974, Fig. 2.32.)

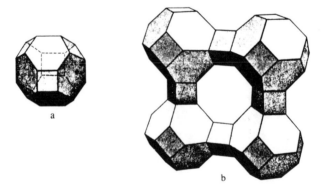

Figure 13.   The truncated octahedron (a) and (b) the array of truncated octahedra in the framework of zeolite A.   The linkage is shown via the double 4-rings. (From Breck, 1974, Fig. 2.33.)

$(Al_4Si_4O_{16})$ in the centers of the edges of a cube of edge length 12.3 Å. This arrangement produces truncated octahedral units centered at the corners of a cube. Thus, each corner of the cube is occupied by a truncated octahedron enclosing a cavity with a free diameter of 6.6 Å. The center of the unit cell is a large cavity which has a free diameter of 11.4 Å, usually referred to as the alpha-cage. The packing of α-cages and a space-filling model of zeolite A are shown in Figure 14. (See also Figure 4a, Chapter 1).

A stereodiagram of the framework topology of zeolite A is shown in Figure 15. Crystal structure data are given in Appendix II, Table 10. Because of its commercial importance, there is a substantial volume of structural data on zeolite A, including details of cations positions for various cationic forms which will not be covered here. However, it is worth noting that the cation exchange of sodium ions with other alkali or alkaline earth ions is used to tailor the adsorption pore size of this zeolite as different cations occupy different positions in the dehydrated structure and, hence, obstruct to some degree the access of the adsorbed molecule to the intercrystalline void through the eight-membered rings.

Group 4 (D6R)

The framework structures of zeolites of Group 4 are characterized by the double 6-ring (D6R) as the secondary building unit. This group includes the minerals chabazite, gmelinite, and faujasite as well as the commercially important synthetic zeolites X and Y which have the same framework topology as faujasite. Synthetic zeolites X and Y are used throughout the world in the catalytic cracking of petroleum to produce gasoline and as catalysts in many other petrochemical processes. Natural faujasite is a rare zeolite, having been found in minute amounts in only a few basalt-vug occurrences.

Faujasite and Faujasite-type Structures. The topology of faujasite and of the related synthetic zeolites X and Y is obtained by linking sodalite units with double 6-rings or hexagonal prisms. Each sodalite unit is linked through hexagonal prisms to four sodalite units in a tetrahedral configuration. A stereodiagram of the aluminosilicate framework of faujasite and synthetic zeolites X and Y is shown in Figure 16. Crystal structure data are listed in Appendix II, Tables 11, 12, and 13. The structure is further depicted in Figure 17 which shows the spatial arrangements of truncated octahedral units in a diamond-like array and a space-filling model showing the approximate location of oxygen ions in the framework. The framework structure of faujasite is very open and encloses a system of large cages linked by four windows of 12-rings to adjacent cavities.

37

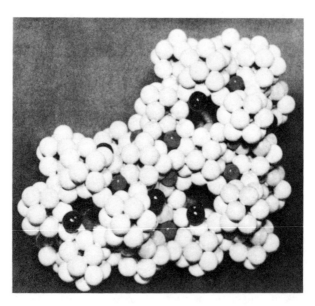

Figure 14. (a) The truncated cuboctahedron and (b) the arrangement in the framework of zeolite A. (c) A space-filling model of the structure of zeolite A showing the packing of oxygen. Three unit cells are shown as are typical locations of cations in site I and II. The site-I cations are indicated by the gray spheres and the site-II cations by the black spheres. A ring of oxygens lies in each cube face. Six apertures open into each α-cage with a free diameter of 4.2 Å. Each α-cage is connected to 8 β-cages by distorted 6-rings with a free diameter of 2.2 Å. The surface of the α-cage is characterized by 8 monovalent ions coordinated in a planar configuration with 6 oxygens of the framework and 4 ions in a one-sided coordination with at least 4 framework oxygens. (From Breck, 1974, Figs. 2.34, 2.36.)

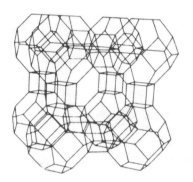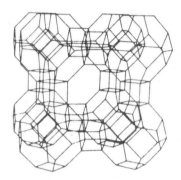

Figure 15.   Stereodiagram of framework topology of zeolite A.   (From Meier and Olson, 1971).

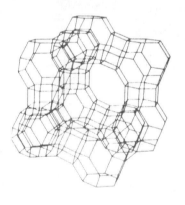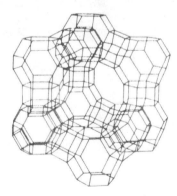

Figure 16.   Stereodiagram of framework topology of faujasite.   (From Meier and Olson, 1971.)

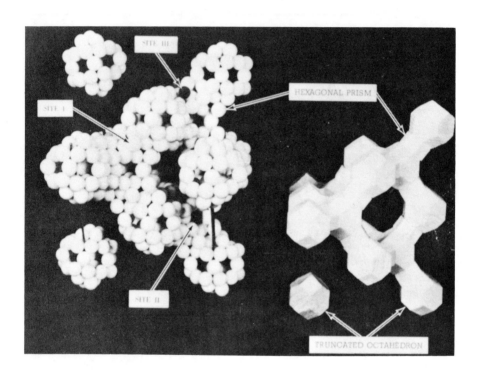

Figure 17. The structure of zeolites X, Y, and faujasite as depicted by a model showing the spatial arrangement of truncated octahedral units in a diamond-type array. The space-filling model showing the approximate location of oxygen atoms in the framework is also pictured. The arrangement of truncated octahedra in one unit cell is shown. The large 12-ring is visible as well as the smaller 6-rings which form the apertures into the β-cages. The cation positions typical of zeolite X are illustrated as site I within the hexagonal prism unit, site II adjacent to the single 6-rings, and site III within the main cavities. (From Breck, 1974, Fig. 2.38.)

There are also extensive structural data published on faujasite and the synthetic zeolites X and Y in terms of the details of cation positions and the location of water molecules. Such data are treated comprehensively by Breck (1974) and Smith (1976).

Chabazite. The aluminosilicate framework of chabazite consists of D6R units arranged in layers in the sequence ABCABC.... A stereogram of the framework topology of chabazite is shown in Figure 18. The D6R units are linked by tilted 4-rings, and the resulting framework contains large, ellipsoidal cavities, each of which is entered by six apertures that are formed by the 8-rings. Crystal structure data for chabazite are listed in Appendix II, Table 14. A space-filling model of chabazite is shown in Figure 9 and in Figure 4b of Chapter 1.

Group 5 (T_5O_{10} Units)

All structures in Group 5 are based on cross-linked chains of tetrahedra. An individual chain is composed of linked units of five tetrahedra designated by Meier (1968) as a 4-1, T_5O_{10} secondary building unit (see Figures 1, 19). The three possible ways of crosslinking the chains produce the three types of framework structures of Group 5. This group includes the minerals natrolite, scolecite, mesolite, thomsonite, gonnardite, and edingtonite. The characteristic needlelike habit of zeolite minerals in this group reflects the common structural characteristic of cross-linked chains of tetrahedra.

Natrolite, Scolecite, Mesolite. Although they differ in unit-cell composition and symmetry, natrolite, scolecite, and mesolite possess the same type of framework structure, illustrated in Figure 19. Crystal structure data for natrolite are listed in Appendix II, Table 15. Natrolite and scolecite are, respectively, the sodium and calcium forms of the same framework types; mesolite is the intermediate member of the group with a sodium/calcium ratio of 1.

Thomsonite, Gonnardite, Edingtonite. The framework structure of thomsonite is based on the second type of crosslinking the chains of 4-1 units and is shown in Figure 20. Crystal structure data for thomsonite and edingtonite are listed in Appendix II, Tables 16 and 17, respectively. Like other zeolites of Group 5, thomsonite occurs as needlelike crystals. Water molecules are located in the channels in double zigzag chains. Gonnardite has a similar aluminosilicate framework, but its Si/Al ratio of 1.5 is somewhat larger than that of thomsonite (Si/Al = 1.0). Edingtonite represents the simplest method of crosslinking the chains of 4-1 units, as illustrated in Figure 21.

Figure 18a. Stereoscopic view of the framework structure of chabazite. This view is
approximately parallel to the rhombohedral axis. The horizontal 6-rings
are linked through tilted 4-rings to other horizontal rings that are displaced both vertically and laterally. The centers of the 6-rings can be
seen to lie in about the same pattern as the packing of spheres in simple
crystals. The cubic and hexagonal close packing of spheres is generally
represented as ABC or AB where A, B, and C represent the three possible
horizontal projections. The same terminology can be applied if we designate the center of the hexagonal ring by the alphabetical symbol. When
adjacent layers have the same letter designation, i.e., AABBCC, the
structure contains D6R units. If the adjacent layers have different
symbols, hexagonal rings are linked by tilted 4-rings. There are many
possible structures that can be built hypothetically from parallel 6-
rings in this way. (From Breck, 1974, Fig. 2.19.)

Figure 18b. A stereoscopic view of the framework of chabazite illustrating the main
adsorption cavity or cage and the approximate location of the three cation
sites. Site I in the center of the double 6-ring is at the origin of the
unit cell and provides for a near octahedron of oxygen atoms. There is
one site-I cation per unit cell. Site II is located near the center of
the 6-ring but displaced away from site I into the main cavity. There
are two site-II cations per unit cell. Site III occurs in 6 pairs of
positions, each pair about 1 Å apart and near the tilted 4-ring. The
locations indicated in the figure are approximate. (From Breck, 1974,
Fig. 2.20.)

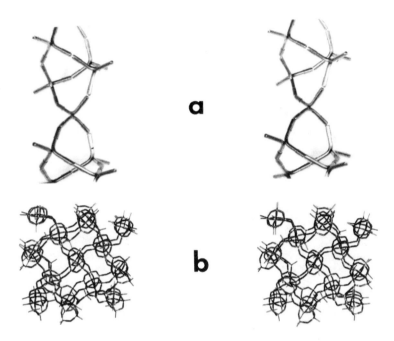

Figure 19. (a) Stereoscopic view of a model of the chain of T_5O_{10} units of tetrahedra as present in the zeolites of Group 5. (b) Stereoscopic view of the framework structure of natrolite. The view is nearly parallel to the c-axis. The tetrahedra chain has a fundamental unit length of 6.6 Å so that one dimension of the unit cell will be 6.6 Å or a multiple. (From Breck, 1974, Fig. 2.24.)

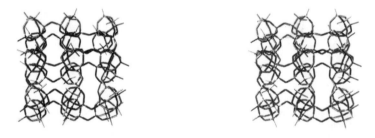

Figure 20. Stereoscopic view of the framework structure of thomsonite nearly parallel to the c-axis. The main channels in thomsonite are perpendicular to the fiber or c-axis. Eight of the 12 cations in the hydrated zeolite are found in coordination with 7 oxygens, 4 of which are framework oxygens and 3 of which are in 8-fold coordination. Si, Al ordering due to Si/Al = 1 requires c = 2 x 6.6 Å or two 4-1 units. (From Breck, 1974, Fig. 2.25.)

43

 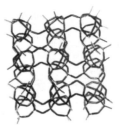

Figure 21. Stereoscopic view of the framework structure of edingtonite parallel to the c-axis. (From Breck, 1974, Fig. 2.26.)

Group 6 (T_8O_{16} Units)

The common structural element in the zeolites of Group 6 is a special configuration of 5-rings shown in Figure 22. This secondary building unit of six tetrahedra was called a 5-1 unit by Meier (1968). These units form complex chains which are linked to each other in various ways, as shown in Figure 23. Group 6 includes the zeolite minerals mordenite, dachiardite, ferrierite, epistilbite, and bikitaite.

Mordenite. Mordenite is the most siliceous zeolite and has a nearly constant Si/Al ratio of 5.0, suggesting an ordered distribution of Si and Al in the framework structure. A stereodiagram of the framework topology of mordenite is shown in Figure 24, and a packed-sphere model showing oxygen ions and channels parallel to the c-axis is shown in Figure 25. Crystal structure data are listed in Appendix II, Table 18. The structure consists of chains of cross-linked 5-rings, a feature that is probably responsible for the high thermal stability displayed by this zeolite. The chains combine to form twisted 12-membered rings which span vertical, near-cylindrical channels. The adsorption properties of natural mordenite are inconsistent with the channel dimensions predicted by this model of 6.7 Å. As only molecules considerably smaller than this value can be adsorbed in dehydrated natural mordenite, the term "small port" mordenite was coined. The smaller adsorption pore size is believed to be due to diffusion blocks produced either by stacking faults in the structure in the c-direction or by the presence of amorphous material within the channels. A synthetic type of mordenite has been prepared which exhibits the adsorption properties characteristic of those expected from the structure (Keough and Sand, 1961). This material is characterized by free diffusion of molecules in the main channel and has been termed "large port" mordenite.

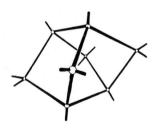

Figure 22. Diagram showing the configuration of the T_8O_{16} units of tetrahedra as found in the framework topology of Group 6 zeolites. (From Breck, 1974, Figure 2.46.)

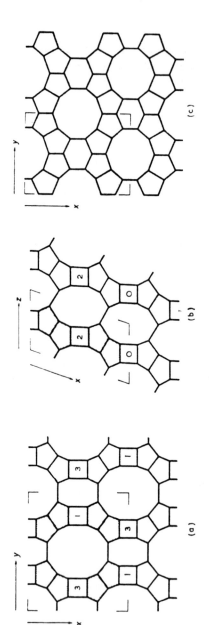

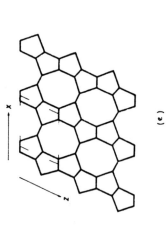

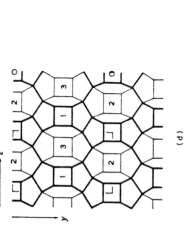

Figure 23. Skeletal structures of mordenite (a), dachiardite (b), ferrierite (c), epi-stilbite (d), and bikitaite (e) projected along the axes of the main channels. The numbers give the heights of the 4-membered rings as multiples of one quarter of the respective cell constant. (From Meier, 1968.)

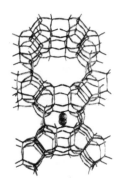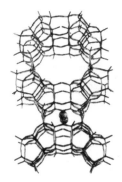

"Molecular Sieve Zeolites - I"

Figure 24. Stereoscopic view of the framework structure of mordenite parallel to the main c-axis showing the large 12-rings and the positions of four of the sodium ions which are located as shown in the constructions with a minimum diameter of 2.8 Å. The remaining four sodium ions per unit cell probably occupy some of the 8- and 12-fold positions at random. (From Breck, 1974, Fig. 2.27.)

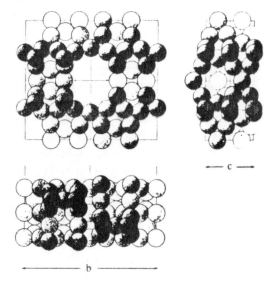

"Molecular Sieve Zeolites - I"

Figure 25. Packed sphere drawing of mordenite showing oxygen atoms and channels parallel to c and parallel to b. The large continuous channels parallel to c are elliptical with dimensions of 6.7 x 7.0 Å. The channels parallel to b consist of pockets, with dimensions of 2.9 x 5.7 Å separated by 2.8 Å restrictions. (From Breck, 1974, Fig. 2.48.)

Dachiardite, Ferrierite. The aluminosilicate frameworks of dachiardite and ferrierite are formed by different types of linking of the cross-linked 5-ring chains found in mordenite. Stereograms of these frameworks are shown in Figures 26 and 27, and crystal structure data are listed in Appendix II, Tables 19 and 20.

Group 7 ($T_{10}O_{20}$ Units)

Zeolites classified in Group 7 include the minerals heulandite, stilbite, brewsterite, and clinoptilolite, all of which occur as platy or lath-shaped crystals. The secondary building unit for Group 7 is the special configuration of tetrahedra shown in Figures 1 and 28 of 4-4-1 or $T_{10}O_{20}$ units. This SBU contains 4- and 5-rings arranged in sheets, thereby accounting for the cleavage properties of the minerals. The different manners of connecting these units are shown in Figure 29.

Heulandite, Clinoptilolite. The arrangement of $T_{10}O_{20}$ units in the framework of heulandite is shown in Figure 29; the framework topology is shown in Figure 30. The low bond density between the layers is readily apparent. Because of the low bond strength in one direction, heulandite changes structurally on dehydration. If dehydrated at moderate temperatures below 130°C, heulandite will adsorb H_2O and NH_3. If it is dehydrated at higher temperatures, no adsorption occurs.

Although clinoptilolite presumably has the same crystal structure as heulandite, it is considerably more stable towards dehydration than heulandite and readily adsorbs H_2O and CO_2. Some varieties adsorb O_2 and N_2. Its chemical composition is significantly different from that of heulandite in Si/Al ratio and exchangeable cations. The thermal stability of clinoptilolite to 700°C is also considerably greater than that of heulandite. Recent incomplete structural data of Chen et al. (1978) suggest that some clinoptilolites from California differ slightly from heulandite in framework topology. The heulandite structure has pores defined by both 8- and 10-membered rings; whereas, the related clinoptilolite structure has pores defined by only 10-membered rings.

Crystal structure data for heulandite and clinoptilolite are listed in Appendix II, Tables 21 and 22.

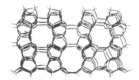 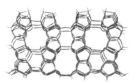

Figure 26.   Stereoscopic view of the framework structure of dachiardite shown parallel
             to the main c-channel. The main channel is parallel to the b-axis [010]
             and intersects channels which parallel the c-axis. The minimum free aper-
             ture diameter of both of these channels is about 4 Å. (From Breck, 1974,
             Fig. 2.28.)

 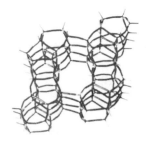

Figure 27.   Stereoscopic view of the framework structure of ferrierite parallel to the
             c-axis. The main c-axis channels are shown in cross section. Two inter-
             secting systems of channels run through the ferrierite structure parallel
             to the b- and c-axes. There is no channel in the a-direction. This is
             also the case with other zeolites in this group. The c-axis channel has
             an elliptical cross-section; dimensions = 4.5 x 5.5 Å and a cross-sectional
             area of about 18 Å^2. The second system of channels, parallel to the b-axis,
             is formed by 8-rings with diameters of 3.4 x 4.8 Å. Located in these
             channels are cavities which are approximately spherical with a diameter of
             about 7 Å. There is only one diffusion path available to a molecule of
             any moderate size; in order to move from one major channel to another,
             the diffusing species must move through the smaller 8-ring aperture.
             (From Breck, 1974, Fig. 2.29.)

Figure 28. Configuration of the $T_{10}O_{20}$ units of tetrahedra in the framework structures of Group 7 zeolites.   (From Breck, 1974, Fig. 2.50.)

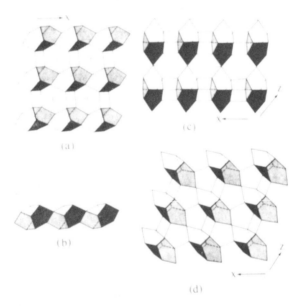

Figure 29. Arrangement of the $T_{10}O_{20}$ units of tetrahedra in the framework structures of (a) brewsterite, (b) brewsterite chain of 4-rings, (c) stilbite, and (d) heulandite. The sheets are parallel to [010] and are connected by relatively few oxygen bridges which results in two-dimensional channel systems. The bond densities across these connecting bridges are very low, 1.7 (Si,Al)-O bonds per 100 Å^2 in heulandite and stilbite compared to 4.2 in analcime. This accounts for the lamellar habit of the crystals. The larger channels, parallel to a, are formed by rings of 10 tetrahedra. (From Breck, 1974, Fig. 2.51.)

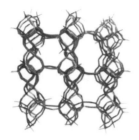 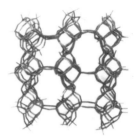

Figure 30. Stereoscopic view of model of the framework structure of heulandite. The
channel system is two dimensional; one set parallel to the a-axis formed
by 8-rings, 4.0 x 5.5 Å in free aperture, and two sets parallel to the
c-axis formed by 8- and 10-rings with free apertures of 4.1 x 4.7 Å and
4.4 x 7.2 Å. The cross-section parallel to the ab plane is similar to
that of mordenite. (From Breck, 1974, Fig. 2.32.)

ACKNOWLEDGMENTS

This chapter has been prepared largely from the comprehensive treatise "Zeolite
Molecular Sieves" by D.W. Breck (1974). The author is indebted to Dr. Breck and Wiley-
Interscience for their permission to reproduce much of the material contained herein.
The invaluable assistance of Dr. Breck and Dr. R.L. Patton, Union Carbide Corporation,
is gratefully acknowledged.

REFERENCES

Breck, D.W. (1974) Zeolite Molecular Sieves: Wiley-Interscience, New York, 771 pp.

Chen, N.Y., Reagan, W.J., Kokotailo, G.T., and Childs, L.P. (1978) A survey of cata-
    lytic properties of North American clinoptilolites: In, Sand, L.B. and Mumpton,
    F.A., Eds., Natural Zeolites: Occurrence, Properties, Use, Pergamon Press,
    Elmsford, New York, 411-420.

Fischer, K.F. and Meier, W.M. (1965) Kristallchemie der zeolithe: Fortschr. Mineral.
    42, 50-86.

Flanigen, E.M., Khatami, H., and Szymanski, H.A. (1971) Infrared structural studies
    of zeolite frameworks: Adv. Chem. Ser. 101, 201-229.

Keough, A.H. and Sand, L.B. (1961) A new intracrystalline catalyst: J. Am. Chem. Soc.
    83, 3536-3537.

Meier, W.M. (1968) Zeolite structures: In, Molecular Sieves, Society of Chemical
    Industry, London, 10-27.

Meier, W.M. and Olson, D.H. (1971) Zeolite frameworks: Adv. Chem. Ser. 101, 155-170.

Smith, J.V. (1976) Structural classification of zeolites: Am. Mineral. Soc. Spec.
    Pap. 1, 281-290.

Smith, J.V. (1976) Origin and structure of zeolites: In, Rabo, J.A., Ed., Zeolite
    Chemistry and Catalysis: ACA Monograph 171, Am. Chem. Soc., Washington, D.C.,
    3-79.

Chapter 3

GEOLOGY OF ZEOLITES IN SEDIMENTARY ROCKS

R. L. Hay

INTRODUCTION

Zeolites occur in rocks of many types and in a wide variety of geologic settings. Most of the earlier geological work on zeolites was carried out on typical occurrences in cavities of basalt flows. In the 1950s, however, careful X-ray diffraction studies of bedded tuffs in the western part of this country and in Japan and Italy showed that larger quantities of zeolites are to be found in sedimentary rocks and that many such strata consist almost entirely of zeolite minerals. As an introduction to the several sections that follow on the occurrence of zeolites, this chapter outlines the geological characteristics of the types of occurrences that have been recognized in sedimentary environments since 1950. No attempt has been made to detail the many occurrences of zeolites in the vugs and cavities of mafic volcanics, as these have been well documented in the geologic literature over the last 200 years.

Mineralogy

Comprehensive listings of zeolites in sedimentary rocks have been given in several review papers, including Coombs et al. (1959), Hay (1966), Iijima and Utada (1972), Sheppard (1971), and Olson et al. (1975), to which the reader is referred. Five zeolites predominate in sedimentary rocks: analcime, clinoptilolite, heulandite, laumontite, and phillipsite. Ranking next in abundance are chabazite, erionite, mordenite, natrolite (with gonnardite), and wairakite. Alkali-rich zeolites include analcime, clinoptilolite, erionite, mordenite, and natrolite; whereas, heulandite, laumontite, and wairakite are calcic. Phillipsite and chabazite range from alkalic to calcic depending on their mode of occurrence, and vary greatly in their $Si/(Al+Fe^{3+})$ ratio (Table 1). In a general way, the degree of hydration of individual species varies inversely with the Si/Al ratio. Also, with the exception of analcime and natrolite, the more siliceous zeolites tend to contain sodium and potassium as exchangeable cations, rather than calcium.

Table 1. Main Compositional Features of Zeolites
Found in Sedimentary Rocks.

Zeolite	$Si/(Al+Fe^{3+})$	Dominant Cations
Clinoptilolite	4.0 - 5.1	K > Na
Mordenite	4.3 - 5.3	Na > K
Heulandite	2.9 - 4.0	Ca,Na
Erionite	3.0 - 3.6	Na,K
Chabazite	1.7 - 3.8	Ca,Na
Phillipsite	1.3 - 3.4	K,Na,Ca
Analcime	1.7 - 2.9	Na
Laumontite	2.0	Ca
Wairakite	2.0	Ca
Natrolite	1.5	Na

Zeolites are formed by reaction of solid materials with pore water. Volcanic glass is a common reactant as are x-ray-amorphous and poorly crystalline clay, montmorillonite, plagioclase, nepheline, biogenic silica, and quartz. Both clay minerals and zeolites can form from the parent material, and whether a clay mineral or zeolite is formed depends on the physical environment and on the activities of dissolved species such as H^+, alkali- and alkaline-earth cations, and H_4SiO_4. Phyllosilicates are generally favored by high activities of Mg^{2+} and by high ratios of H^+ to Na^+, K^+, and Ca^{2+}. The species of zeolite which crystallizes will depend on the temperature, pressure, and activities of various ions and on the activity or partial pressure of H_2O. It should be noted that the chemical potential of water can be lowered and the less hydrous zeolites favored by increasing the temperature, the ionic strength, or the partial pressure of carbon dioxide (p_{CO2}). These matters are discussed more fully elsewhere (for example, Miyashiro and Shido, 1970).

Early-formed zeolites commonly react with pore fluid to yield other zeolites in sedimentary rocks. This can result from changes in the physical or chemical environment, or simply because sufficient time has elapsed for a less-stable form to transform to a more stable one. Phillipsite, mordenite, and clinoptilolite, for example, can be replaced by analcime, and analcime by laumontite.

Occurrences

Most zeolite occurrences in sedimentary rocks can be categorized into several types of geologic environments or hydrologic systems, including (1) saline, alkaline lakes; (2) saline, alkaline soils and land surfaces; (3) deep-sea sediments; (4) open hydrologic systems; (5) hydrothermal alteration zones; and (6) burial diagenetic or metamorphic environments. Many of these types of occurrence result in characteristic patterns of zeolite zoning which are diagrammatically illustrated in Figures 1 and 2. Zoning in saline-lake deposits is exemplified by mineral assemblages in altered vitric tuffs. It can be traced horizontally for several kilometers and probably reflects salinity gradients in the original lake water. In saline, alkaline surface soils, zeolites are formed at and near the land surface. The marine-sediment type is based

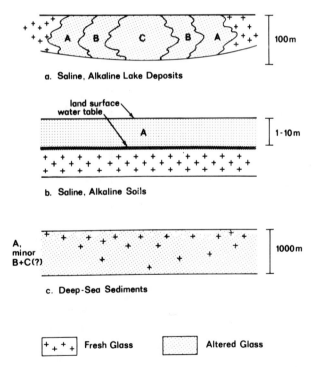

a. Saline, Alkaline Lake Deposits

b. Saline, Alkaline Soils

c. Deep-Sea Sediments

Fresh Glass   Altered Glass

Figure 1. Diagrams showing patterns of authigenic zeolites and feldspars in tuffs of saline, alkaline lakes; saline, alkaline soils; and deep-sea sediments. Zone A is characterized by non-analcimic, alkali-rich zeolites, zone B by analcime, and zone C by feldspars.

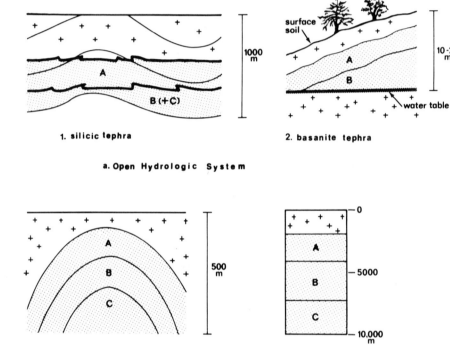

Figure 2.   Diagrams showing patterns of authigenic zeolites and feldspars in tuffs
where the zonation is of (a) open-system type, (b) hydrothermal, and (c)
a result of burial diagenesis.   Zone A is characterized by non-analcimic
alkali-rich zeolites, zone B by analcime or heulandite, and zone C by
K-feldspar in (a_1) and by albite with or without laumontite in (b) and (c).
Symbols are the same as in Figure 1.   (From Hay, 1978.)

on the occurrence of zeolites in deep-sea sediments, which commonly increase in abundance as a function of depth and age, at least in tuffaceous deposits. The open-system type refers to accumulations of reactive tephra up to several hundred meters thick which show a vertical zoning of authigenic, silicate minerals, largely attributable to the progressive changing of meteoric water by reaction with solid material along its flow path. Hydrothermal alteration zones are those produced by local, abnormally high geothermal gradients, as in areas of volcanic activity, and are commonly associated with hot springs. Burial diagenesis refers to the vertical zonation of authigenic minerals in thick, sedimentary accumulations attributable chiefly to the increase of temperature with depth. Obviously, these categories are not mutually exclusive, and burial diagenesis can modify assemblages produced by other mechanisms.

SALINE, ALKALINE-LAKE DEPOSITS

Zeolites are both common and widespread in deposits of saline, alkaline lakes, and the purest concentrations of zeolites are found in such lacustrine tuffs. Saline, alkaline lakes are found in arid and semiarid regions and have a pH of about 9.5 as a result of dissolved sodium carbonate-bicarbonate or sodium borate. A wide variety of materials react to yield zeolites in this alkaline environment, reflecting the solubility of aluminosilicate detritus at high pH. Among the more common reactants are volcanic glass, biogenic silica, poorly crystalline clay, montmorillonite, kaolinite, plagioclase, and quartz. In addition to zeolites, sodium silicates, K-feldspar, and borosilicates also form in this environment. Reactions are relatively rapid, and vitric tuff can alter to zeolites in about 1,000 years.

The most common zeolites formed from silicic glass are phillipsite, clinoptilolite, and erionite; less common are mordenite and chabazite. Near-monomineralic beds of clinoptilolite, phillipsite and erionite replacing rhyolitic tuffs are widespread in Plio-Pleistocene lake deposits of the western United States (Sheppard, 1971). In water of moderately high salinity these zeolites can alter to analcime; in highly saline water, all zeolites can alter to potassium feldspar. A typical alteration sequence might be glass altering to a zeolite such as clinoptilolite, which may alter to analcime, which in turn may be replaced by K-feldspar (Hay, 1966; Sheppard and Gude, 1968). Tuff beds in saline-lake deposits more than a few hundred thousand years old may therefore show a mineralogic zoning: (a) an outer zone of glass either unaltered or replaced by clay minerals where the lake water was relatively fresh; (b) a zeolitic zone of alteration to clinoptilolite, phillipsite, etc. where the water

was saline; (c) a more saline zone characterized by analcime; and (d) a highly saline
zone in the center of the basin where tuffs may be altered largely to K-feldspar
(Figure 1). Sheppard and Gude (1968, 1973) described excellent examples in late
Cenozoic rocks; the Green River Formation is an Eocene example (Hay, 1966; Surdam
and Parker, 1972). Analcime is generally the only zeolite found in non-tuffaceous
claystones interbedded with zeolitic tuffs containing phillipsite or clinoptilolite.
Like zeolites in tuffs, it readily reacts to form K-feldspar in the highly saline zone.

SOILS AND SURFACE DEPOSITS

Zeolites are formed readily from suitable materials at the land surface where the
pH is high as a result of the concentration of sodium carbonate-bicarbonate by evapo-
transpiration in an arid or semiarid climate. A wide variety of zeolites have formed
in "soils" formed from both tuffaceous and non-tuffaceous sediments of Pleistocene
and Holocene age in the region of Olduvai Gorge, Tanzania (Hay, 1976). This is a
semiarid region which now, and in the past, has been characterized by high concentrations of sodium carbonate-bicarbonate at the land surface. Wind-worked, or eolian
tuffs, mostly of nephelinite composition, were altered to phillipsite, chabazite,
natrolite, and analcime during deposition to form a well-cemented rock. In such
environments, volcanic glass reacts at about the same rate as in saline, alkaline
lakes (Hay, 1976). Stream-worked trachytic tuffs were altered to zeolites to a
maximum depth of 18 m beneath one of these units of eolian tuff. The lower surface
of zeolitic alteration probably represents the water table, and the stream-worked
tuffs between the water table and the eolian tuffs were altered by sodium carbonate
solutions periodically flushed downward from the surface during rains. Non-tuffaceous
fluvial claystones were altered on floodplains penecontemporaneous with deposition to
form analcime and lesser amounts of other zeolites. Claystones richest in zeolites
(15-40%) are bright reddish-brown and may represent a relatively modern, small-scale
example of Mesozoic "analcimolites," which are widespread, reddish-brown, terrestrial
deposits rich in analcime (Hay, 1966).

DEEP-SEA DEPOSITS

Zeolites have formed widely at relatively low temperatures in marine sediments.
Most of our information about marine zeolites comes from studies of sea-floor sediments, in particular the cores obtained by the Deep Sea Drilling Project (DSPS).

Phillipsite and clinoptilolite are the dominant zeolites, analcime is next in abundance, and erionite, natrolite, and mordenite occur rarely. Authigenic minerals associated with zeolites include smectites, palygorskite, sepiolite, cristobalite, and quartz. Zeolites are particularly abundant in volcaniclastic sediments, especially vitric ash, where they may form as much as 80% of the layer. As might be expected, phillipsite is commonly associated with low-silica, generally basaltic, tephra, and clinoptilolite with siliceous tephra. However, phillipsite can be the principal or the only zeolite in silicic ashes, and clinoptilolite can be the zeolite of mafic ash layers. Analcime is found principally in mafic volcanic deposits. Clinoptilolite is widespread and can be a major component of pelagic detrital clays with biogenic silica. Small amounts of clinoptilolite are found in most other types of fine-grained marine sediments, as for example, siliceous ooze, carbonate ooze, and green clay. Clinoptilolite is much more common than phillipsite in the Atlantic than in the Indian and Pacific Oceans, where phillipsite predominates. Within the Indian and Pacific Oceans, clinoptilolite becomes more abundant at depth and is the principal zeolite in Cretaceous rocks.

Small amounts of zeolite are found at shallow depth in Quaternary sediments, indicating that they can form in substantially less than a million years. Czynscinski (1973) inferred from studies of Indian Ocean sediments that phillipsite crystals can grow to their full size ($\sim$45 μm) in as little as 150,000 years. Although rapid by marine standards, this is far slower than in saline, alkaline lakes. Overall, the content of zeolites increases as a function of age, at least in tuffaceous sediments, reflecting the progressive alteration of volcanic glass. However, some unaltered glass can be found in sediments as old as Cretaceous.

Much, if not most, of the zeolites in sea-floor sediments was formed by reaction of glass with pore water, with or without the addition of biogenic silica. Smectite and palygorskite are associated with zeolites in altered tephra, and the amount of authigenic clay minerals commonly exceeds that of zeolites. Zeolites are also formed by reaction of poorly crystalline or x-ray-amorphous clays. Biogenic silica contributes to the formation of clinoptilolite, and clinoptilolite casts of radiolarians and and forams have been widely noted in non-tuffaceous sediments. Pore fluid of the DSDP cores generally varies rather little in composition from sea water except for silica, which commonly ranges from about 5 to 65 ppm. The content of dissolved silica bears a highly variable but probably significant relation to the nature of the zeolites. Pore waters of cliniptilolite-bearing sediments have, on the average, a higher content of SiO_2 than is found in the waters of phillipsite-bearing sediments.

Causes of the distribution of clinoptilolite and phillipsite are controversial, and represent a major unresolved problem in the zeolites of marine sediments.

OPEN-SYSTEM TYPE DEPOSITS

In open hydrologic systems, tephra sequences commonly show a more or less vertical zonation of zeolites and other authigenic minerals, reflecting the chemical change in meteoric water as it moved through the system. Clay minerals, most commonly smectites, are formed by hydrolysis of vitric ash in the upper part of the system, increasing the pH and dissolved solids to the point where glass alters chiefly to zeolites. Flow in an open hydrologic system is either downward or has a downward component where meteoric water enters the system, resulting in a vertical to gently inclined zonation of water composition and authigenic minerals.

Thick, nonmarine accumulations of silicic tephra may contain an upper zone, 200 to 500 m thick, containing fresh glass, montmorillonite, and opal. The next lower zone, as much as 500 or more meters thick, is generally characterized by a siliceous zeolite such as clinoptilolite. An underlying zone may be characterized by analcime, with or without authigenic K-feldspar and quartz (Figure 2). The John Day Formation of Central Oregon (Oligo-Miocene) is an excellent example (Hay, 1963). Alkali-rich, low-silica tuffs may exhibit a comparable zonation of zeolites with thicknesses on the order of ten meters. Koko Crater, Hawaii, and the Neapolitan Yellow Tuff of Naples are examples (Hay and Iijima, 1968; Sersale, 1958). This zonation at relatively shallow depths reflects the highly reactive nature of alkali-rich, silica-poor glass.

HYDROTHERMAL DEPOSITS

Zeolites are widespread in areas of hydrothermal alteration and may also exhibit a zonal pattern. Clinoptilolite or mordenite characterizes the shallowest and coolest zones while progressively deeper zones commonly contain analcime or heulandite, laumontite, and wairakite. Well-known examples are in Yellowstone Park, Wyoming; Wairakei, New Zealand; and Onikobe, Japan. The mineralogic zonation represents primarily the progressive dehydration with increasing temperature.

Alteration zones surrounding Kuroko-type ore deposits in Japan exhibit a complex zonation resulting from a combination of submarine hydrothermal alteration and low-temperature burial diagenesis (Iijima, 1974). In the simplest form, an analcime zone extends laterally from a phyllosilicate hydrothermal zone and is progressively bordered by mordenite and clinoptilolite-mordenite zones.

Curiously, laumontite is readily formed by hydrothermal alteration, yet apparently has not been synthesized in unseeded hydrothermal experiments from anything but dehydrated laumontite (Coombs et al., 1959). Nature's ease in forming laumontite is illustrated by the laumontite precipitated in radiators fed by natural hot spring water in Pauzhetka City, Kamchatka (Lebedev and Gorokhova, 1968)!

BURIAL DIAGENETIC DEPOSITS

This type, also termed burial-metamorphic, comprises those deposits which were formed over wide areas in thick accumulations of volcanoclastic sediments at significantly increased temperatures. Deposits up to 12 km thick are known which display vertical zonation of zeolites and associated minerals. In Cenozoic sequences, a surface zone as much as 2 km thick may contain fresh glass, and successively lower zones may have assemblages with (1) mordenite and clinoptilolite, (2) analcime and heulandite, and (3) laumontite and albite. The laumontite-albite zone grades into a zone with prehnite and pumpellyite, representing a transition into the greenschist facies of regional metamorphism. Locally, a wairakite-bearing zone may be intercalated between rocks containing laumontite and those with prehnite and pumpellyite. In pre-Cenozoic strata, the uppermost zones have generally been eroded away. The mineralogic zones represent a decrease in hydration with depth caused by an increase in temperature. As might be expected, zeolitic assemblages correlate rather well with coal rank, which has been shown to be directly related to burial depth and geothermal gradient.

Probably the most instructive examples of burial-diagenetic zoning are in late Cenozoic volcaniclastic sediments of the Green Tuff region of Japan (Iijima and Utada, 1972; Utada, 1971). Here, the thickness and mineralogy of the zones can be correlated to some extent with the geothermal gradient, and zeolitic diagenesis is presently proceeding in some areas, allowing an estimate of the temperatures at which the zones are forming. The best-known example is the Niigata oil field, where drill core is available for a 5-km thickness of marine strata (Iijima and Utada, 1971). An uppermost zone, 0.9-1.9 km thick, contains fresh glass, and an underlying zone, 1.6-2.5 km thick, contains mordenite and clinoptilolite. Successively lower are an analcimic zone, about 1 km thick, and a zone with albite. Analcime was formed by reaction of clinoptilolite and mordenite, as shown by pseudomorphs, and albite was formed by reaction of quartz and analcime. Temperatures at the top of the zeolitic rocks are

61

41-49°C, and those of the analcime-albite transition are 120-124°C. This temperature is significantly lower than the 190°C inferred on the basis of laboratory experiments.

The upper zones of burial diagenesis are mineralogically similar to those of the open hydrologic system, and the two types can be difficult to distinguish in thick sequences of silicic tephra. Indeed, the distinction between the two types is to some extent artificial as an increase in temperature with burial will speed up open-system reactions, and water-rock interactions can be expected to change the composition of the pore fluid in any thick sequence of tuffaceous deposits. There are, however, fundamental differences in the two types which may be recognizable. In the open-system type, a sharp contact may separate the fresh-glass zone from the clinoptilolite zone, whereas in burial diagenesis the contact is gradational, and a thick transitional zone may be present. In the open-system type, zeolitic alteration is generally short-lived and penecontemporaneous with deposition, and it may not correspond to the time of deepest burial.

ZEOLITES IN IGNEOUS ROCKS

Most of the zeolite literature prior to 1960 deals with the zeolites in amygdules and fillings of other cavities in igneous rocks, particularly basic lavas. Low-silica lavas commonly contain a variety of zeolites, mostly the low-silica types, and not rarely several species are found in a single amygdule. Silicic lavas and ignimbrites are characterized by a relatively small number of silicic zeolites, the most common of which are mordenite and clinoptilolite.

Walker (1960) documented in a 10,000-m thickness of Cenozoic lavas in Iceland a zonal distribution of zeolites comparable to that in thick sequences of tuffaceous sediments. The zeolite zones are flat lying, cut across stratification, and are approximately parallel to the original top of the lava pile. The uppermost zone lacks zeolites, and the lower zones represent basalt-water reactions at increasing burial depth and temperature. This zoned sequence appears to be an example of burial diagenesis, and the zeolites in most lava sequences buried more than about 1 km are probably burial diagenetic. Another, spectacular example is that of the superb zeolite crystals in the Triassic lavas of New Jersey (Schaller, 1932). Other zeolite occurrences are restricted to the vicinity of hot springs or vent areas of volcanoes and are the result of localized hydrothermal alteration of lava.

Analcime and other zeolites are common in many alkaline, low-silica lavas that have neither been buried nor hydrothermally altered. The zeolites may occur in the groundmass and in veins and cavity fillings. The origin of these is uncertain, and they may either have crystallized deuterically or later, at low temperature, by reaction of meteoric water with the lava. The origin of analcime "phenocrysts" in alkaline lavas is similarly controversial. It is not presently clear whether these minerals crystallized in a melt or formed after eruption by low-temperature replacement of leucite.

REFERENCES

Coombs, D.S., Ellis, A.J., Fyfe, W.S., and Taylor, A.M. (1959) The zeolite facies, with comments on the interpretation of hydrothermal syntheses: Geochim. Cosmochim. Acta 17, 53-107.

Czyscinski, K. (1973) Authigenic phillipsite formation rates in the central Indian Ocean and the equatorial Indian Ocean: Deep-Sea Res. 20, 555-559.

Hay, R.L. (1963) Stratigraphy and zeolitic diagenesis of the John Day Formation of Oregon: Univ. Calif. Pubs. Geol. Sci. 42, 199-262.

Hay, R.L. (1966) Zeolites and zeolitic reactions in sedimentary rocks: Geol. Soc. Am. Spec. Paper 85, 130 pp.

Hay, R.L. (1976) Geology of the Olduvai Gorge: Univ. Calif. Press, Berkeley, 203 pp.

Hay, R.L. (1978) Geologic occurrence of zeolites: In, Sand, L.B. and Mumpton, F.A., Eds., Natural Zeolites: Occurrence, Properties, Use, Pergamon Press, Elmsford, New York, 135-143.

Hay, R.L., and Iijima, A. (1968) Nature and origin of palagonite tuffs of the Honolulu Group on Oahu, Hawaii: Geol. Soc. Am. Mem. 116, 331-376.

Iijima, A. (1974) Clay and zeolitic alteration zones surrounding Kuroko deposits in the Hokuroku District, northern Akita, as submarine hydrothermal-diagenetic alteration products: In, Ishihara, S., Kanehira, K., Sasaki, A., Sato, T., and Shimazaki, Y., Eds., Geology of Kuroko Deposits, Soc. Mining Geol. Japan, Tokyo, 267-289.

Iijima, A., and Utada, M. (1971) Present-day zeolitic diagenesis of the Neogene Geosynclinal deposits in the Niigata oil field, Japan: In, Advances in Chemistry Series 101, Molecular Sieve Zeolites - I, Am. Chem. Soc., 342-349.

Iijima, A., and Utada, M. (1972) A critical review on the occurrence of zeolites in sedimentary rocks in Japan: Jap. J. Geol. Geog. 42, 61-84.

Lebedev, L.M., and Gorokhova, L.M. (1968) Recent mineralization in technical structures in Pauzhetka City (Kamchatka): Dokl. Akad. Nauk S.S.S.R. 182, 1399-1401.

Miyashiro, A., and Shido, F. (1970) Progressive metamorphism in zeolite assemblages: Lithos 3, 251-260.

Olson, R.H., Breck, D.W., Sheppard, R.A., and Mumpton, F.A. (1975) Zeolites: In, Lefond, S.J., Ed., Industrial Minerals and Rocks, Am. Inst. Min. Met. Pet. Engin., New York, 1235-1274.

Schaller, W.T. (1932) The crystal cavities of the New Jersey zeolite region: U.S. Geol. Surv. Bull. 832, 90 pp.

Sersale, R. (1958) Genesi e costituzione del tufo giallo napoletano: Rend. Accad. Sci. Fis. Mat. 25, 181-207.

Sheppard, R.A., and Gude, A.J., 3d (1968) Distribution and genesis of authigenic silicate minerals in tuffs of Pleistocene Lake Tecopa, Inyo County, California: U.S. Geol. Surv. Prof. Paper 597, 38 pp.

Sheppard, R.A. and Gude, A.J., 3rd (1973) Zeolites and associated authigenic silicate minerals in tuffaceous rocks of the Big Sandy Formation, Mohave County, Arizona: U.S. Geol. Surv. Prof. Pap. 830, 36 pp.

Surdam, R.C., and Parker, R.D. (1972) Authigenic aluminosilicate minerals in the tuffaceous rocks of the Green River Formation, Wyoming: Geol. Soc. Am. Bull. 83, 689-700.

Utada, M. (1971) Zeolitic zoning in the Neogene pyroclastic rocks of Japan: Sci. Papers Coll. Gen. Educ. Univ. Tokyo 21, 189-221.

Walker, G.P.L. (1960) Zeolite zones and dike distribution in relation to the structure of the basalts of eastern Iceland: J. Geol. 68, 515-528.

Chapter 4

ZEOLITES IN CLOSED HYDROLOGIC SYSTEMS

Ronald C. Surdam

INTRODUCTION

Although zeolite minerals were identified in saline, aklaine-lake deposits in
the western United States in the 1920s (e.g., Ross, 1928; Bradley, 1928), it was not
until the late 1950s that such occurrences were found to be commonplace in the volcanic regions of this and other countries. It is now apparent that zeolites are
among the most common authigenic silicates in sedimentary rocks and that they occur
in rocks of diverse age, lithology, and depositional environment (Hay, 1966; Sheppard,
1971; 1973; Mumpton, 1973b; Munson and Sheppard, 1974). Several classifications of
zeolite deposits in sedimentary rocks have been offered, and although they differ
in details, all agree that a principal type includes those deposits formed in closed
hydrologic basins (Hay, 1966; Sheppard, 1973; Mumpton, 1973b; Kossovskaya, 1975).
Such deposits generally result from the reaction of volcanic glass with connate
water trapped during sedimentation in saline, alkaline lakes. From the large number
of zeolitic rocks that have formed in this manner, it is obvious that this type of
depositional environment is an excellent habitat for zeolite growth. The combination
of highly reactive pyroclastic material and saline, alkaline solutions is apparently
ideal for the crystallization of zeolites.

Of the near 40 known zeolite species, only six are common in saline, alkaline-
lake deposits, namely analcime, chabazite, clinoptilolite, erionite, mordenite, and
phillipsite. These six minerals show considerable range in their Si:Al ratio and in
their exchangable cations and water content. The distribution of individual species
in saline alkaline-lake deposits and the relationship of zeolite phases to other
authigenic silicates, such as smectite, potassium feldspar, and searlesite, are
still open questions, although the careful work of Sheppard and Gude (1968, 1969, 1973)
documenting the lateral zonation of zeolites in three closed hydrologic basins of
California and Arizona has laid a firm foundation for our understanding of

such deposits. Paragenesis considerations of zeolite deposits in closed hydrologic basins have also been aided by the existence of several modern saline, alkaline lakes where zeolites are presently forming. The importance of solution chemistry in zeolite formation has been discussed by Surdam and Eugster (1976) with reference to several of these deposits in the Lake Magadi region of East Africa.

For an interpretation of zeolite deposits in closed hydrologic basins, it is essential to examine not only the mineralogy of the deposits, but also the geology, hydrology, and chemistry of the basins. With this background many of the mineralogical observations that were previously not understood become clarified. This chapter therefore deals with the geological, hydrological, and chemical processes responsible for the formation of zeolites in saline, alkaline-lake deposits.

GEOLOGIC SETTING

Modern Closed Basins

Nearly all modern, closed hydrologic basins are in one of two tectonic settings: (1) block-faulted regions, e.g., the Basin and Range province of the western United States, or (2) trough valleys associated with rifting, e.g., the East Rift Valley of Kenya. Schematic cross section of the two types of closed basins is shown in Figure 1. These examples are idealized and it is possible to develop closed basins which are combinations of the two end-member types. One type, designated a "playa-lake complex," is a broad flat valley surrounded by high mountain ranges. The other, designated a "rift system," is a steep walled, flat, narrow valley. An important difference between these two is the geohydrology. In the playa-lake complex, a perennial playa lake occupies the lowest portion of the valley floor. The ground water flow is generally shallow, and much of it will discharge at the mud-flat or playa fringe adjacent to the lake (Figure 1A). The ground water circulation in the rift system, however, is deep, and discharge is by springs along faults at the edge of the lake (Figure 1B).

An important feature of both types is that only a limited amount of clastic debris reaches the lake. In a playa lake, the sharp break in slope at the foot of the mountains traps most of the clastic material in alluvial fans, and most streams flowing beyond the fans are intermittent. In a rift valley, most of the flow into the lake is by springs which do not carry clastic debris. During dilute or humid stages, or perhaps early in the evolution of the basin, perennial streams or rivers

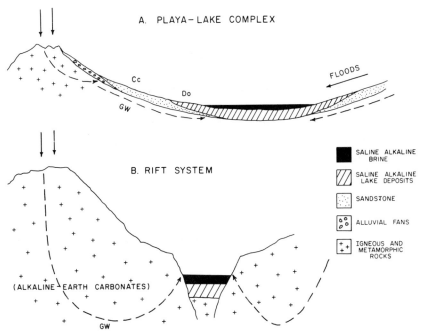

Figure 1. Hydrology and brine evolution in a playa-lake complex (A) and a rift
system (B). Solid arrows, rainwater; GW, ground water circulation path;
Cc, calcite precipitation; Do, dolomite formation. (Modified from
Eugster and Surdam, 1973.)

may reach the valley floor, such as those which feed Pyramid and Walker Lakes in Nevada.
Perennial streams may transport a relatively large supply of clastic material; however, in arid regions the clastic debris that reaches the valley floor is generally
transported by sheet-flow processes during storms.

Deep Springs Lake in eastern California and Teels Marsh in western Nevada are
excellent examples of the playa-lake type of closed hydrologic basin (Jones, 1966;
Smith and Drever, 1976). Lake Magadi in Kenya is an outstanding example of the
rift-valley type (Eugster, 1970). In both types, significant systematic hydro-
chemical changes occur, particularly from the point of ground water discharge to the
center of the basin.

Hydrographic Features

A lake in a closed basin is a dynamic feature because the area of the lake, the
depth of water, and the salinity vary greatly according to seasonal inflow and

evaporation (Langbein, 1961). Thus, fluctuations in lake level and salinity are the result of a delicate imbalance between precipitation and evaporation. All lakes in closed hydrologic basins are characterized by such fluctuations due to seasonal or short-term climatic variations. There are also many examples of longer term variations in closed basins. The level of Lake Rudolf, Kenya, has fallen approximately 46 m in the last 3,000 years as a result of climatic factors (Fuchs, 1939). During the period 1904 to 1930, the net decrease in lake level averaged 23-30 cm/yr. However, the imbalance between inflow and evaporation is a delicate one, and Fuchs estimated that an increase of 12.7 cm/yr in the rate of precipitation would be sufficient to cause a rise in the level of Lake Rudolf. The most significant control on the existence of closed lakes is evaporation (Langbein, 1961).

Lakes lacking outlets or lakes in closed basins are exclusive features of the arid and semi-arid zones where annual evaporation exceeds rainfall. Russell (1896, p. 131-132) concluded: "The study of the present geography of the earth shows that in regions where the mean annual precipitation exceeds 20 or perhaps 25 inches, inclosed lakes do not occur, although the topographic conditions may be favorable." Langbein (1961) interpreted this phenomenon as follows: "The greater the aridity, the greater is the rate of net evaporation. The greater the net evaporation, the smaller the area of a closed lake (larger the playa fringe). The smaller the lake area, the more numerous are the topographic opportunities for a closed lake." Thus, saline, alkaline lakes are dependent not only on a special geological setting but also on an arid or semi-arid climate.

Brine Evolution

From the point of view of zeolite genesis, a significant characteristic of a closed hydrologic basin is the brine evolution, an understanding of which simplifies the explanation of mineral patterns. Saline, alkaline brines, in contrast to merely saline brines, ideally develop in a closed hydrologic basin as follows: (1) Soft water forms by reaction of CO_2-charged ground water with igneous and metamorphic rocks (Jones, 1966; Garrels and MacKenzie, 1967; Hardie and Eugster, 1970) and emerges at the foot of alluvial fans. The ground water flow from the alluvial fans to the center of the basin is shallow. (2) Where the climate is arid or where dry cycles are characterized by marked seasonal contrasts, evaporative concentration is intense, and calcite precipitates from ground water or soil water in the capillary zone. Both pH and the Mg:Ca ratio increase steadily as the ground water moves toward the center of the basin while evaporation, evaporative pumping, and recycling

68

of efflorescent crusts continues. (3) After the Mg:Ca ratio reaches a value 12, protodolomite forms, and the salinity of the ground water continues to increase until the water reaches the lake at the center of the basin (Eugster and Surdam, 1973). The lake water is not only saline, but also alkaline. In this context, alkalinity is defined (cf. Stumm and Morgan, 1970) as:

$$\text{alkalinity} = HCO_3^- + 2CO_3^= + OH^- - H^+.$$
$$\text{total} \quad \text{total}$$

(4) If evaporation remains intense, a sodium carbonate mineral such as trona eventually precipitates.

In contrast, alkaline-earth playa flats are not present in rift valleys because of differences in hydrology. Here the ground water circulation is very deep; and the lake is fed by alkaline hot springs devoid of Ca and Mg, the precipitation of which must have occurred earlier during underground circulation (Eugster, 1970).

A second very important difference is that in a playa-lake complex, the shallow ground water is characterized by systematic chemical changes from the point of groundwater discharge, across the mud flats, and to the lake at the center of the basin. Generally, the change is from fresh water to saline, alkaline brines. The ground water in a rift system, however, is not characterized by shallow circulation; and when it emerges at the lake margin, it is in the form of concentrated saline, alkaline waters (Eugster, 1970). Thus, the shallow sediments in a playa-lake complex are subjected to ground waters that vary systematically from dilute water to concentrated brines; whereas, sediments in a rift system are subjected only to concentrated spring water and brines.

Weathering Reactions

As Hardie and Eugster (1970) suggested, the compositional history of closed-basin waters can be separated into two phases: (1) acquisition of solutes by the dilute waters through weathering-type reactions with soil and bedrock, and (2) evaporative concentration which eventually causes the precipitation of very soluble minerals. Jones (1966) suggested that the composition of saline, alkaline waters in closed basins is chiefly the result of weathering processes accompanied by simple evaporative concentration and selective mineral precipitation. For example, in the absence of older chemogenic deposits, water compositions are related to the hydrolysis of silicates by CO_2-charged waters, as shown schematically below:

$$\text{Primary silicate} + H_2O + CO_2 = \text{clay} + Ca^{+2} + Na^+ + HCO_3^- + SiO_2.$$

The primary silicates react with CO_2-charged water, i.e., soil water, and a clay mineral is formed that remains in the soil profile. Cations, HCO_3^-, and SiO_2 are released into solution. Garrels and MacKenzie (1967) have shown that if this type of water is isolated and concentrated by isothermal evaporation in equilibrium with the atmosphere, it produces a highly alkaline $Na-HCO_3-CO_3$ water.

Chemical Trends

To produce saline, alkaline brine by evaporation, the parent waters are severely restricted in terms of possible compositional variations. The parent waters must be rich in HCO_3^-, the $HCO_3^-/(Ca^{+2} + Mg^{+2})$ molar ratio must be distinctly greater than unity, and other anions such as Cl^- and SO_4^{-2} must be minor (for details, see Hardie and Eugster, 1970). This is the type of water derived as a result of the weathering of igneous or metamorphic rocks (Garrels and MacKenzie, 1967). However, the presence of abundant pyritic shales or older evaporites in the drainage basin precludes the development of alkaline brines (Jones, 1966).

During evaporation, the chemical evolution of the dilute waters can be understood in terms of "chemical divides." The most important chemical divide in the evolution of alkaline brines is the early precipitation of alkaline earth carbonates. This divide determines the major compositional trend that a water follows and is determined in the evaporation history, commonly by calcite precipitation (Hardie and Eugster, 1970). If alkalinity/$2(mCa^{+2}) > 1$, the solution will be depleted in Ca^{+2} by calcite precipitation and will with further evaporation be enriched in CO_3^{-2} by calcite precipitation. Further evaporation will cause enrichment in Ca^{+2} and will result in a saline, but not alkaline lake, such as the Great Salt Lake. In Figure 2, the open circles represent the composition of Green River water (alkalinity/$2(mCa^{+2})$ > 1). If evaporation occurs, calcite will be the first mineral to precipitate, and CO_3^{-2}-rich water (and ultimately an alkaline brine) will result. The black circle in Figure 2 represents the composition of the Jordan River which provides 20% of the inflow to the Great Salt Lake (alkalinity/$2(mCa^{+2})$ < 1). With evaporation, calcite again will be the first mineral to precipitate; however, a Ca^{+2}-rich water and ultimately a saline brine will result. Evaporation of any dilute water whose composition falls above the dahsed line (alkalinity = $2Ca^{+2}$) will result in a saline, alkaline brine, but evaporation of water whose composition plots below the dashed line will result in a saline, but not alkaline brine.

Other important chemical divides are summarized in Figure 3 (modified after Hardie and Eugster, 1970). The ratio of alkality/alkaline earth ions determines the

70

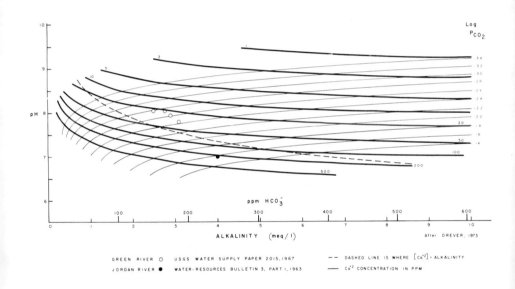

Figure 2. Graphical display of one of the most significant chemical divides in brine evolution. The dashed line is where Ca^{+2} exactly balances alkalinity. Water plotting above dashed line, upon further concentration will precipitate calcite and evolve into an alkaline brine. Water plotting below the dashed line, upon precipitating calcite, will evolve into a saline but not an alkaline brine. (After Drever, 1972).

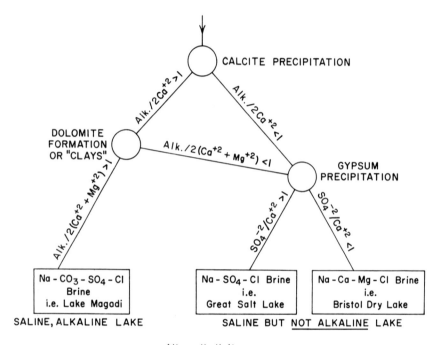

CALCITE PRECIPITATION

$Alk./2Ca^{+2} > 1$

$Alk./2Ca^{+2} < 1$

DOLOMITE FORMATION OR "CLAYS"

$Alk./2(Ca^{+2} + Mg^{+2}) < 1$

GYPSUM PRECIPITATION

$Alk./2(Ca^{+2} + Mg^{+2}) > 1$

$SO_4^{-2}/Ca^{+2} > 1$

$SO_4^{-2}/Ca^{+2} < 1$

Na - CO$_3$ - SO$_4$ - Cl
Brine
i.e. Lake Magadi

Na - SO$_4$ - Cl Brine
i.e.
Great Salt Lake

Na - Ca - Mg - Cl Brine
i.e.
Bristol Dry Lake

SALINE, ALKALINE LAKE

SALINE BUT NOT ALKALINE LAKE

Alk. = alkalinity

Figure 3. Flow sheet for evaporative concentration of dilute waters. (Modified from Hardie and Eugster, 1970.)

DEEP SPRINGS PLAYA - LAKE

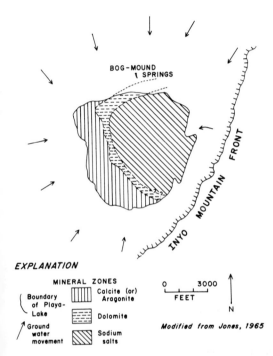

Figure 4. Distribution of mineral zones
at modern Deep Springs playa, California
(Jones, 1965). Playa-lake deposits surrounded by marginal silts and sands.

EXPLANATION

MINERAL ZONES

Boundary of Playa-Lake	▥ Calcite (or) Aragonite
Ground water movement	▤ Dolomite
	▨ Sodium salts

0   3000
FEET

N

Modified from Jones, 1965

compositional trend early in the evaporation history of the water (ionic
strength < 0.1) and is controlled by the
kinds of rocks exposed undergoing weathering in the hydrologic basin. For alkaline brines to develop, the hydrologic
basin must be characterized by igneous
or metamorphic rocks. If large quantities of pyritic shales or older evaporites are exposed in the basin, the al-
kalinity/$2(Ca^{+2}+Mg^{+2})$ ratio will be less
than unity and will preclude the formation of alkaline brines. Thus, a
closed basin surrounded by igneous and
metamorphic rocks, or by sediments other
than pyritic shales and evaporites, is
an ideal setting to develop saline, alkaline brines and lakes.

As already discussed, in a closed basin precipitation of alkaline earth carbonates profoundly affects the composition of the waters and hence the sequence and
nature of the saline minerals formed in the center of the basin. The result is the
formation of a mineral zonation with the most soluble minerals occupying the center
of the basin and segregated from the less soluble phases. Jones (1965) and Hardie
(1968) documented this type of mineral zonation at Deep Springs Valley and Saline
Valley in eastern California (Figure 4).

Efflorescent Crusts

Another important process in the formation of concentration brines in closed
basins is the recycling of efflorescent crusts. The enrichment of NaCl in brines

associated with playa lakes is the result of occasional rains that dissolve the most soluble material from the efflorescent crust covering the playa surface, and subsequent flood water or ground water that transport the solutes to the adjacent lacustrine environment. Hardie (1968) showed that efflorescent crusts of modern playa surfaces are typically halite because the crusts are products of complete dehydration of brine drawn to the surface by capillary action. Recently, it has been recognized that resolution of efflorescent crusts is a significant method of increasing the salinity of ground water and lacustrine water in closed basins (Jones et al., 1976).

Volcanism

Volcanism commonly accompanies block faulting or rifting. For example, both the Basin and Range province of the western United States and the East Rift Valley of Kenya contain abundant volcanic rocks. According to Maxey (1968), Tertiary and Quaternary volcanic rocks cover about 30% of the state of Nevada; whereas, sediments of the same age, which are in part of volcanic origin, cover 48%. Baker (1958; 1963) showed that nearly all the Tertiary and Quaternary stratigraphic section in the East Rift Valley of Kenya consists of volcanic rocks. A tectonic setting of block or rift faulting is therefore characterized by an abundance of glassy, pyroclastic material, ranging in composition from intermediate to silicic. Much of the volcanism in the East Rift Valley resulted in trachytic pyroclastic material; whereas much of the volcanism in the Basin and Range Province resulted in rhyolitic pyroclastic material.

MINERAL PATTERNS

Lacustrine deposits that contain zeolites commonly show the following geological features: (1) fossil assemblages indicative of lacustrine environments, i.e., "freshwater" diatoms, ostracodes, turtles, and fish; (2) laminated mudstone that interfingers laterally with fluviatile sandstone and conglomerate containing vertebrate fossils; (3) non-marine evaporites; (4) "chemical deltas" and tufa or spring deposits; and (5) strand-line deposits such as algal stromatolites, ostracode lag deposits, ooliths and pisoliths, flat-pebble conglomerates, and beach sands.

These features strongly suggest that such lacustrine deposits were, at least in part, saline and alkaline. The most obvious indication is the occurrence of saline minerals. Some deposits containing zeolites also contain bedded saline minerals, such as trona, nahcolite, and halite; whereas, others contain disseminated crystal

74

molds of gaylussite or pirrsonite. Dolomitic mudstone is a further indication of
saline, alkaline conditions (Jones, 1965). The distribution of fossils also indicates
abnormal salinity.   Freshwater diatoms and ostracodes are commonly found in sediments
near inlets, but away from these areas, there is an absence of fossils.

Indirect evidence of alkalinity is the abundant calcite at the lake margin. The
calcite is best explained as precipitates formed where inflowing calcium-bearing "fresh"
water mixed with alkaline water. This phenomenon can be observed at modern Mono Lake
in California and at Pyramid Lake in Nevada. Smith (1966) described similar chemical
deltas of calcium carbonate in the upper Wisconsin sediments at the inlet of Searles
Lake.

The most distinguishing feature of the saline-lake zeolite deposits is the lateral
zonation of minerals. Zeolite deposits of the hydrothermal, burial metamorphic, and
open-system types commonly show a vertical mineral zonation but not a lateral zonation
(Sheppard, 1975). Recent studies on the distribution of authigenic silicate minerals
in silicic tuffs of Cenozoic closed-basin deposits have shown a consistent pattern.
In a general way, there is a lateral gradation basinward of fresh glass to zeolites
to potassium feldspar. The tuffs of Pleistocene Lake Tepoca in southern California
exemplify this pattern (Figure 5) (Sheppard and Gude, 1968). Fresh glass occurs along
the margin and at inlets of the ancient lake. The glass is succeeded inwardly by a
zone of zeolites and in the central part of the lake basin, by potassium feldspar.
The zeolites in the Lake Tepoca deposit are chiefly phillipsite, erionite, and clinoptilolite. It should be noted that in most cases the tuff beds have been altered almost
completely to zeolites or potassium feldspar. Horizons containing more than 90% of a
single zeolite are not uncommon.

A variation in the distribution pattern of authigenic silicate minerals in tuffs
of closed-basin deposits is illustrated by the Big Sandy Formation of northwestern
Arizona, a series of lacustrine strata that were deposited in a Pliocene saline,
alkaline lake near Wikieup. A non-analcimic zeolite facies occurs along the margin
of the ancient lake deposit and is succeeded basinward by an analcime facies followed
by a potassium-feldspar facies. Zeolites in the non-analcimic zeolite facies are
chiefly chabazite, clinoptilolite, erionite, and phillipsite. A similar distribution
pattern has been recognized in the Miocene Barstow Formation of southern California
(Sheppard and Gude, 1969) and in the Eocene Green River Formation of Wyoming (Surdam
and Parker, 1972).

In addition to zeolites, potassium feldspar, clay minerals, and alkaline-earth
carbonates, saline, alkaline-lake deposits locally contain opal or chalcedony, quartz,
searlesite ($NaBSi_2O_6 \cdot H_2O$), fluorite (CaF_2), and dawsonite ($NaAl(CO_3)(OH)_2$) of

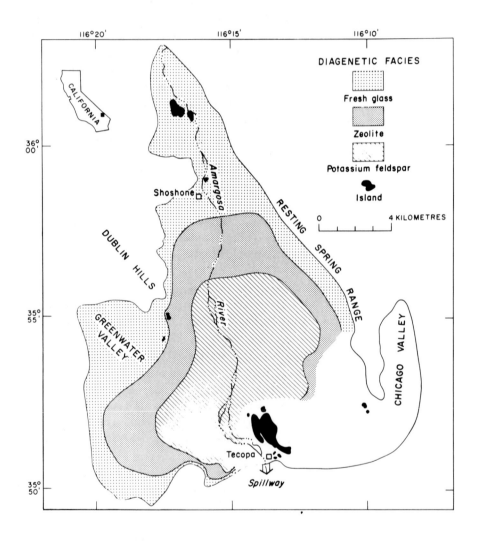

Figure 5.   Map of ancient Lake Tecopa near Shoshone, California, showing diagenetic facies for a composite of all the tuff beds in the deposit. (After Sheppard and Gude, 1968.)

authigenic origin. Saline minerals commonly associated with these deposits are trona $(Na_2CO_3 \cdot NaHCO_3 \cdot 2H_2O)$, nahcolite $(NaHCO_3)$, pirssonite $(Na_2CO_3 \cdot CaCO_3 \cdot 2H_2O)$, gaylussite $(Na_2CO_3 \cdot CaCO_3 \cdot 5H_2O)$, and halite $(NaCl)$.

Pleistocene Lake Tecopa Deposits

Three diagenetic mineral facies are recognized in the tuffs of the Lake Tecopa deposits. Tuffs nearest the lake margin are characterized by fresh glass and are termed the "fresh-glass facies." Tuffs in the central part of the lake basin are characterized by potassium feldspar and/or searlesite and are termed the "potassium feldspar facies." Those tuffs between the fresh-glass facies and the potassium feldspar facies and characterized by zeolite minerals are termed the "zeolite facies."

The boundaries between the facies are laterally gradational and difficult to recognize in the field. Tuffs of the fresh-glass facies generally can be distinguished from those of the zeolite facies in the field, but tuffs of the zeolite facies almost never can be distinguished from those of the potassium feldspar facies (Sheppard and Gude, 1968). X-ray diffractometer data of bulk samples, coupled with thin section study, are considered essential for positive identification and placement in the proper facies.

Figure 5 shows the distribution of the mineral facies in the Pleistocene Lake Tecopa deposits (Sheppard and Gude, 1968). It is important to note that the pattern shown is for a composite of all tuff beds. The distribution of potassium feldspar and searlesite is similar throughout the basin. In a general way, the facies boundaries parallel the shape of the lake basin in plan (see Figure 5). The zeolite facies is generally 0.8-1.6 km wide except near the Amargosa River inlet at the north end of the basin, where it is broadened to 2.0-2.8 km, suggesting that the chemical deposition environment affected, if not controlled, the formation of specific minerals.

Pliocene Big Sandy Formation

Three diagenetic mineral facies are recognized in the tuffs of the Pliocene Big Sandy Formation in Arizona. Tuffs nearest the margin of the formation are characterized by zeolites other than analcime and are termed the "non-analcimic zeolite facies" (Sheppard and Gude, 1973). Tuffs in the central part of the lake basin are characterized by potassium feldspar and are termed the "potassium feldspar facies." Those tuffs between the non-analcimic zeolite facies and the potassium feldspar facies are characterized by analcime and are called the "analcime facies" (Sheppard and Gude, 1973). No relict fresh glass was recognized in any of the tuffaceous rocks of the Big Sandy Formation.

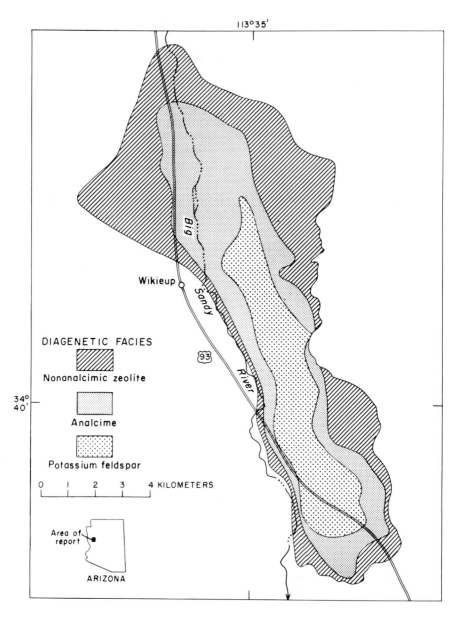

Figure 6. Map of the Big Sandy Formation near Wikieup, Arizona, showing diagenetic facies for a composite of all the tuffaceous rocks. (After Sheppard and Gude, 1973.)

78

The boundaries between the Big Sandy facies are also laterally gradational and difficult to recognize in the field. Figure 6 shows the diagenetic mineral facies for a composite of all the tuffaceous rocks in the Big Sandy Formation. This same pattern is observed in individual tuff beds (Sheppard and Gude, 1973).

The mineral facies in plan are elongated parallel to the depositional basin. The non-analcimic zeolite facies and the analcime facies are broadest at the northern part of the basin which was the major inlet of the ancient lake. The non-analcimic zeolite facies is also broad and extends basinward near the other inlets. In a general way, the non-analcimic zeolite facies and the analcime facies narrow where the basin narrows. About 5 km southeast of Wikieup, where the basin was narrowest, both the non-analcimic zeolite and analcime facies are absent along the eastern margin (Figure 6). Again, these features suggest that the chemical depositional environment affected, if not controlled, the distribution of the diagenetic mineral facies.

Miocene Barstow Formation and Eocene Green River Formation

Two other examples of lateral zonation of diagenetic mineral facies in zeolitic closed-basin deposits are the Miocene Barstow Formation in California (Sheppard and Gude, 1969) and the Eocene Green River Formation in Wyoming (Surdam and Parker, 1972). Both have a zone of analcime separating the other zeolites from a zone of potassium feldspar. Minor relict fresh glass occurs in the Barstow Formation, but none has been found in the Green River Formation. The zeolite facies of the Green River Formation grades laterally toward the basin margin into a clay facies (Surdam and Parker, 1972).

The sediments of Lake Tecopa, the Big Sandy Formation, the Barstow Formation, and the Green River Formation range in age from Pleistocene to Eocene and have several features in common: (1) They were subjected to very shallow burial and show only slight deformation; (2) exposures are good, and tuffs can be traced throughout most of the basins, (3) they show an abundance and variety of authigenic silicate minerals in the tuffs; and (4) they show a pronounced lateral zonation of diagenetic minerals, summarized in Figure 7.

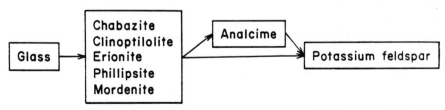

Figure 7. Schematic reaction sequence that leads to lateral mineral zonation that characterizes many saline, alkaline-lake deposits. (After Sheppard and Gude, 1973).

INTERPRETATION OF MINERAL PATTERNS

The boundaries of the above-mentioned mineral zones can be described by the following reactions:

1. glass → alkalic, silicic zeolites
2. alkalic, silicic zeolites → analcime
3. analcime → K-feldspar
4. alkalic, silicic zeolites → K-feldspar

A determination of the chemical parameters that significantly affect these reactions is therefore important to an understanding of the genesis of diagenetic mineral zones in saline, alkaline-lake deposits.

Formation of Alkalic, Silicic Zeolites from Silicic Glass

Evidence supporting the reaction of glass to zeolites is unequivocal (Figure 8). Not only is there a reciprocal relation between glass and zeolites in vitric tuffs containing zeolites, but commonly the zeolites are pseudomorphous[1] after glass shards (Figure 9). Deffeyes (1959) hypothesized that zeolites form during diagenesis by solution of the shards and subsequent precipitation of zeolites from the solution, rather than by "devitrification" of the shards. This mechanism was supported by scanning electron microscope studies by Mumpton (1973). Surdam and Eugster (1976) demonstrated that in the Lake Magadi region of Kenya erionite can form from trachytic glass by the addition of only H_2O. They also emphasized that the water in contact with the glass during the reaction was an alkaline brine. Hay (1966) showed that alkaline brines are ideal agents for the hydration and solution of volcanic glass. Mariner and Surdam (1970) suggested that an aluminosilicate gel may first precipitate from the solution and then zeolites grow from the gel. A gel phase was recognized in Holocene tuffs at Teels Marsh, Nevada, associated with phillipsite and rhyolitic glass (Surdam and Mariner, 1971).

An individual tuff bed will commonly contain several zeolites that differ significantly in composition, suggesting that either initial compositional inhomogenities existed, or more probable, that compositional parameters changed during the hydration and solution of the glass. Table 1 suggests that the important compositional parameters in zeolite genesis are cation ratios, Si/Al, and H_2O. All of these parameters

----

[1]The term pseudomorphous is used in the petrographic sense in that clusters of individual crystals mimic the shape of pre-existing glass shards.

Figure 8. Scanning electron micrograph showing pitted glass fragment altered to acicular erionite in an ash bed from Jersey Valley, Nevada. (From Mumpton, 1973a.)

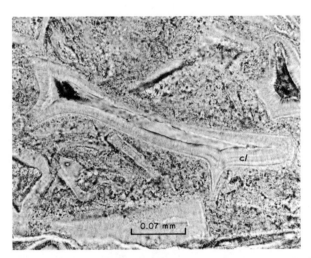

Figure 9. Photomicrograph of zeolitic tuff, showing pseudomorphs of clinoptilolite after shards. The bands in the clinoptilolite are due to variations in the index of refraction along the length of the fibers. The large pseudomorph in center of the photograph has a hollow interior. (After Sheppard and Gude, 1969.)

Table 1. Formulas of Zeolites Found in Saline, Alkaline-lake Deposits.

Name	Dominant Cations	Formula[1]
Analcime	Na	$NaAlSi_{2.0-2.8}O_{6.0-7.6} \cdot 1.0-1.3H_2O$
Chabazite	Na, Ca	$NaAlSi_{1.7-3.8}O_{5.4-9.6} \cdot 2.7-4.1H_2O$
Clinoptilolite	Na, K, Ca	$NaAlSi_{4.2-5.0}O_{10.4-12.0} \cdot 3.5-4.0H_2O$
Mordenite	Na, Ca, K	$NaAlSi_{4.5-5.0}O_{11.0-12.0} \cdot 3.2-3.5H_2O$
Erionite	Na, K	$NaAlSi_{3.0-3.7}O_{8.0-9.4} \cdot 3.0-3.4H_2O$
Phillipsite	Na, K	$NaAlSi_{1.3-3.4}O_{4.6-8.8} \cdot 1.7-3.3H_2O$

[1]Formulas are standardized in terms of a sodium end member that has one aluminum atom. (After Sheppard and Gude, 1969.)

will be affected by changes in salinity and/or alkalinity. Experimental work by Mariner and Surdam (1970) indicated that the Si/Al ratio of zeolites formed by the hydration and solution of glass is related to the pH of the solution. Hay (1966) showed that the rate of solution of silicic glass increases with increasing salinity and alkalinity. The activity of H_2O is related directly to salinity, ranging from 1.0 in distilled water to about 0.7 in the most concentrated modern saline, alkaline brines. The affect of alkalinity on the activity of the alkaline earths was previously demonstrated, i.e., the activity of the alkaline earths decreases rapidly with increasing alkalinity. Thus, the important chemical parameters during the glass-to-zeolite reactions are cation ratios, Si/Al, and activity of H_2O, whereas the most significant parameters affecting the solution of glass are salinity and pH.

Reaction of Alkalic, Silicic Zeolites to Analcime

Analcime is a common constituent of saline, alkaline-lake deposits (Hay, 1966; Sheppard and Gude, 1969, 1973; Hay, 1970; Surdam and Parker, 1972; Surdam and Eugster, 1976). The absence of coexisting analcime and glass in tuffaceous sediments strongly supports the conclusions that analcime does not form directly from the alteration of glass. Analcime is, however, commonly associated with a variety of alkalic, silicic zeolites (Figure 10). Sheppard and Gude (1969, 1973) presented petrographic evidence

showing that clinoptilolite and phillipsite had altered to analcime, and Surdam and
Eugster (1976) documented the reaction of erionite to analcime. Thus, analcime in
tuffs of saline, alkaline-lake deposits formed from precursor zeolites and not
directly from silicic glass.

Sheppard and Gude (1969), studying the diagenesis of tuffs in the Barstow Formation, and Surdam and Eugster (1976), studying the mineral reactions in tuffaceous
rocks of the Lake Magadi region, suggested a correlation between the composition of
the zeolite precursor and the Si/Al ratio of analcime. They observed that analcime
associated with and derived from clinoptilolite and erionite has a relatively high
number of silicon atoms per unit cell (34.5-35.3), whereas analcime associated with
and derived from phillipsite (a less siliceous zeolite) has a low number of silicon
atoms per unit cell (33.1-34.4). Boles (1971) studied the experimental reaction of
clinoptilolite and heulandite to analcime and showed that the Si/Al ratio of the
analcime product is chiefly a function of the Si/Al ratio of the zeolite reactant.
From both field observations and laboratory experiments, there is a strong support
for the reaction of zeolite to analcime; the analcime inherits its Si/Al ratio from
the zeolite precursor.

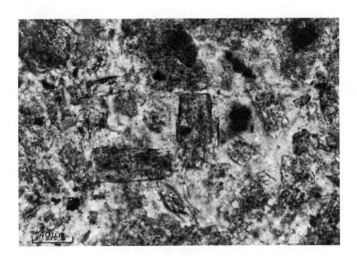

Figure 10. Analcime pseudomorphs after prismatic clinoptilolite. Light areas are
            quartz. (After Sheppard and Gude, 1973).

The reaction of silicic, alkalic zeolites to analcime can be illustrated by the following equations:

Clinoptilolite   Analcime

(1) $Na_2K_2CaAl_6Si_{30}O_{72} \cdot 24H_2O + 4Na^+ \rightarrow 6NaAlSi_2O_6 \cdot H_2O + 18H_2O + 18SiO_2 + 2K^+ + Ca^{+2}$

Phillipsite   Analcime

(2) $NaK_2Al_4Si_{12}O_{32} \cdot 12H_2O + 2Na^+ \rightarrow 4NaAlSi_2O_6 \cdot H_2O + 8H_2O + 4SiO_2 + 2K^+$

Erionite   Analcime

(3) $NaKAl_2Si_7O_{18} \cdot 6H_2O + Na^+ \rightarrow 2NaAlSi_2O_6 \cdot H_2O + 4H_2O + 3SiO_2 + K^+$

The equations show that the activities of the cations, silica, and water control the reactions. The major changes are a gain of sodium and losses of K^+, Ca^{+2}, SiO_2, and H_2O. Furthermore, experimental work and theoretical considerations indicate that the reaction of an alkalic, silicic zeolite to form analcime is favored by a high Na^+/H^+ ratio (Hess, 1966; Boles, 1971), a relatively low activity of SiO_2 (Coombs et al., 1959) or a relatively low Si/Al ratio (Senderov, 1963), and a relatively low activity of H_2O in the pore fluid. Increased salinity and alkalinity favor the formation of analcime because an increase of these parameters decreases the activity of H_2O and the Si/Al ratio, generally affects the cation ratios, and specifically increases the Na^+/H^+ ratio. The above arguments are based on the assumption that equilibrium chemical factors alone are responsible for the formation of analcime; however, kinetic factors may be equally important.

Reaction of Zeolites to Potassium Feldspar

Formation of potassium feldspar from zeolite precursors in tuffs that were never deeply buried has been well documented (Figure 11). Analcime is replaced by potassium feldspar in tuffs of the Eocene Green River Formation, Wyoming (Iijima and Hay, 1968; Surdam and Parker, 1972). Analcime and clinoptilolite are replaced by potassium feldspar in tuffs of the Miocene Barstow Formation, California (Sheppard and Gude, 1969). Potassium feldspar replaced analcime, clinoptilolite, and phillipsite in the Pliocene Big Sandy Formation, Arizona (Sheppard and Gude, 1973). Phillipsite is replaced by potassium feldspar in tuffs of Pleistocene Lake Tecopa, California (Sheppard and Gude, 1968). Potassium feldspar has never been reported to occur with glass, thereby suggesting that it is not a direct product of glass alteration. The

Figure 11. Scanning electron micrograph of large analcime crystals in a matrix of micrometer-sized crystals of potassium feldspar from a tuff near Barstow, California. (From Mumpton and Ormsby, 1976.)

formation of potassium feldspar in these examples has been correlated with a highly saline and alkaline depositional environment; the potassium feldspar is commonly associated with bedded saline minerals.

The reaction of zeolites to potassium feldspar can be illustrated by the following equations:

Analcime   K-feldspar

$$(4) \quad NaAlSi_2O_6 \cdot H_2O + SiO_2 + K^+ \quad \rightarrow \quad KAlSi_3O_8 + H_2O + Na^+$$

Phillipsite   K-feldspar

$$(5) \quad Na_2K_2Al_4Si_{12}O_{32} \cdot 12H_2O + 2K^+ \quad \rightarrow \quad 4KAlSi_3O_8 + 12H_2O + 2Na^+$$

Clinoptilolite   K-feldspar

$$(6) \quad Na_2K_2CaAl_6Si_{30}O_{72} \cdot 24H_2O + 4K^+ \quad \rightarrow \quad 6KAlSi_3O_8 + 24H_2O + 12SiO_2 + 2Na^+ + Ca^{+2}$$

These equations show that the activities of the cations, silica, and water control the reactions. Specifically, the major changes resulting from the reactions are a gain of potassium and losses of sodium and water. Depending on the composition of the zeolite, SiO_2 for the reaction may be in excess or deficient in the zeolite precursor.

Saline and alkaline pore water trapped in the tuffs during deposition, or which evolved during sedimentation and burial, are significant factors responsible for the formation of potassium feldspar. Eugster (1970) showed that the K^+ concentration of brines is highest in the most saline and alkaline brines. Coupled with a high pH, a high potassium concentration would result in a relatively high K^+/H^+ ratio suitable for the crystallization of potassium feldspar. A high pH would also result in increased solubility of SiO_2 and contribute to the stabilization of K-feldspar relative to analcime. A relatively high salinity would lower the activity of H_2O and favor the formation of anhydrous potassium feldspar from hydrous zeolites, including analcime.

Kinetic factors may be important in the zeolite to potassium feldspar reaction. Potassium feldspar was synthesized from natural clinoptilolite in a saturated solution of KOH at 80°C after 44 hours (Sheppard and Gude, 1969). Although this experiment does not duplicate the chemical environment prevalent during the diagenesis of saline, alkaline-lake deposits, it demonstrates the rapidity of the reaction of a zeolite to potassium feldspar in a favorable chemical environment.

Hydrochemistry

The above discussion demonstrates that the chemistry of the pore water is a controlling factor in the mineral reactions that characterize saline, alkaline-lake deposits. Furthermore, a general trend toward increasing salinity and alkalinity favors the glass to zeolite to potassium feldspar reaction sequence.

The relation of brine evolution in closed hydrologic basins to the mineral patterns in saline-alkaline-lake deposits is direct. Brine evolution is gradual and results in systematic chemical gradients. Consequently, in the closed hydrology, the gradients are reflected in centripetal mineral patterns (Figure 4). Figure 12 is a north-south cross section through Deep Springs Valley, California which summarizes the chemical evolution that takes place in a typical closed basin. It should also be noted that from the point of recharge to the saline mineral pan, the Na^+ varies from about 12 to 100,000 ppm, and that K^+ varies from about 5 to 18,000 ppm. The Ca^{+2} concentration, however, varies from about 40 to 3 ppm. Eugster (1970) showed that from the point of recharge to the interstitial brines in the saline mineral pan at Lake Magadi, the SiO_2 concentration varies from 25 to 1,000 ppm or greater. In a closed basin, the concentration factor from dilute inflow in the recharge area to interstitial brines in the discharge area can be as high as 5000. Salinity gradients across the playa lake are also common.

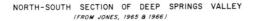

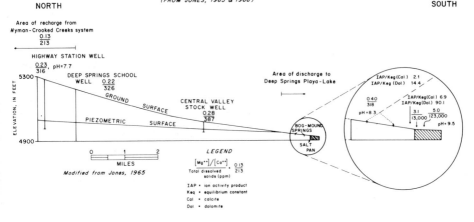

Figure 12. Section showing hydrochemistry of a modern playa-lake complex, Deep Springs Valley, California. Note evolution of brines from north to south as they migrate from alluvial deposits to playa. Diagram constructed from Jones (1965, 1966).

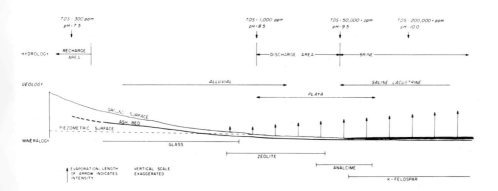

Figure 13. Schematic diagram of the relationships among hydrology, geology, and authigenic mineral zones in a playa-lake complex. The length of the evaporation arrow is proportional not only to the intensity of evaporation but also to the magnitude of the salinity and alkalinity of the pore water. (From Surdam and Sheppard, 1978.)

A probable solution to the problem of diagenetic silicate mineral patterns in tuffaceous rocks in closed basins is suggested by placing a tuff in the context of the hydrochemistry already discussed (Figure 12). A significant fact is that the tuff will act as an aquifer because it is interbedded with relatively impermeable mudrock. As long as the tuff is above the piezometric surface and interbedded with permeable alluvial deposits, it will undergo very little alteration. However, once the ash bed intersects the piezometric surface, it is susceptible to hydration and alteration. This alteration will be pronounced where the ash bed enters the zone of capillary draw, evaporative pumping, and recycling of efflorescent crusts. At this position, salinity and alkalinity of the pore water will increase significantly. This process of concentrating the pore water becomes even more pronounced across the playa from the point of ground water discharge toward the lacustrine environment. The ash bed will be subjected to consistently high salinity and aklalinity in the saline-lacustrine environment because it is there that the effects of evaporation are most pronounced.

Figure 13 is an interpretation of how the hydrochemistry and authigenic mineral patterns of saline, alkaline-lake deposits are related. In the alluvial zone above the water table, the ash will remain relatively unaltered. Near the point of ground-water discharge, particularly in the playa zone, pore water in the ash bed is subjected to capillary draw, and the alteration begins by hydration and solution of glass and eventually the precipitation of alkalic, silicic zeolites. Further concentration of the pore water due to evaporation and recycling of efflorescent crusts occurs near the interface of the playa and the saline, lacustrine environment, and the alkalic, silicic zeolites may react to analcime. In the saline, lacustrine setting, where evaporative concentration is most intense, the salinity and alkalinity are highest, and the alkalic, silicic zeolites or analcime may react to potassium-feldspar. This does not mean that all the mineral reactions take place at the surface, because many reactions take place only after burial. However, Figure 13 demonstrates the development of centripetal salinity and alkalinity gradients in the pore water of a closed basin and also relates the subsequent authigenic silicate minerals to the hydrochemistry.

The closed hydrologic basin setting is a dynamic one, and the position of litho-logic, hydrologic, and mineralogic boundaries will vary with time, mainly as a function of climatic parameters. The dynamic nature of the hydrochemistry explains the fact that almost all the boundaries between authigenic mineral zones are gradational.

Hay (1966) observed in the field, and Mariner (1971) showed experimentally, that silicate-mineral reaction rates increase with increasing salinity and alkalinity.

These data suggest that kinetic factors amplify the silicate mineral zones resulting from the centripetal salinity and alkalinity gradients. On the outer fringe of the playa-alluvial environment, where pore waters have not been greatly concentrated, ash beds suffer only incipient zeolitization. However, that same ash deposited in a saline, alkaline lake containing concentrated pore water undergoes a sequence of authigenic reactions culminating with the formation of potassium feldspar.

Not all authigenic silicate minerals in saline, alkaline-lake deposits are characterized by pronounced centripetal patterns. The Pleistocene High Magadi Beds of Kenya, for example, have no lateral variations in the distribution of authigenic minerals. This is because the lake in which these sediments were deposited was in a narrow fault trough, similar to the present Lake Magadi (Surdam and Eugster, 1976). The wide playa flats that are necessary to produce the lateral zonation cannot develop under these conditions. Therefore, rift-system deposits commonly do not show well-developed lateral mineral variations.

REFERENCES

Baker, B.H. (1958) Geology of the Magadi area: Geol. Survey Kenya Rept. 42, 81 pp.

Baker, B.H. (1963) Geology of the area south of Magadi: Geol. Survey Kenya Rept. 61, 27 pp.

Boles, J.R. (1971) Synthesis of analcime from natural heulandite and clinoptilolite: Am. Mineral. 56, 1724-1734.

Bradley, W.H. (1928) Zeolite beds in the Green River Formation: Science 67, 73-74.

Coombs, D.S., Ellis, A.J., Fyfe, W.S., and Taylor, A.M. (1959) The zeolite facies, with comments on the interpretation of hydrothermal synthesis: Geochim. Cosmochim. Acta 17, 53-107.

Deffeyes, K.S. (1959) Zeolites in sedimentary rocks: J. Petrol. 29, 602-609.

Drever, J.I. (1972) Relations among pH, carbon dioxide pressure, alkalinity, and calcium concentration in waters saturated with respect to calcite at 25°C and one atmosphere total pressure: Contrib. Geol. 11, 41-42.

Eugster, H.P. (1970) Chemistry and origin of the brines of Lake Magadi, Kenya: Mineral. Soc. Am. Spec. Paper 3, 215-235.

Eugster, H.P., and Hardie, L.A. (1975) Sedimentation in an ancient playa-lake complex: the Wilkins Peak Member of the Green River Formation of Wyoming: Geol. Soc. Am. Bull. 86, 319-334.

Eugster, H.P., and Surdam, R.C. (1973) Depositional environment of the Green River Formation of Wyoming: a preliminary report: Geol. Soc. Am. Bull. 84, 1115-1120.

Fuchs, V.E. (1939) The geological history of the Lake Rudolf basin Kenya colony: Royal Soc. London Phil. Trans. 229, 219-274.

Garrels, R.M., and MacKenzie, F.T. (1967) Origin of the chemical composition of some springs and lakes: Advances in Chemistry Series 67, Am. Chem. Soc. 222-242.

Hardie, L.A. (1968) The origin of the Recent non-marine evaporite deposit of Saline Valley, Inyo County, California: Geochim. Cosmochim. Acta 32, 1279-1301.

Hardie, L.A., and Eugster, H.P. (1970) The evolution of closed-basin brines: Mineral. Soc. Am. Spec. Paper 3, 273-290.

Harding, S.T. (1965) Recent variations in the water supply of the Great Basin: Calif. Univ. Water Res. Center Archives, Ser. Rept. 16.

Hay, R.L. (1966) Zeolites and zeolitic reactions in sedimentary rocks: Geol. Soc. Am. Spec. Paper 85, 130 pp.

Hay, R.L. (1970) Silicate reactions in three lithofacies of a semi-arid basin, Olduvai Gorge, Tanzania: Mineral. Soc. Am. Spec. Paper 3, 237-255.

Hess, P.C. (1966) Phase equilibria of some minerals in the $K_2O-Na_2O-Al_2O_3-SiO_2-H_2O$ system at 25°C and 1 atmosphere: Am. J. Sci. 264, 289-309.

Hunt, C.B. (1960) The Death Valley salt pan, a study of evaporite: U. S. Geol. Surv. Prof. Paper 400-B, B456-B457.

Iijima, A., and Hay, R.L. (1968) Analcime composition in the Green River Formation of Wyoming: Am. Mineral. 53, 184-200.

Jones, B.F. (1965) The hydrology and mineralogy of Deep Springs Lake, Inyo County, California: U. S. Geol. Surv. Prof. Paper 502-A, 56 pp.

Jones, B.F. (1966) Geochemical evolution of closed basin waters in the western Great Basin: Symp. Salt, 2d, Ohio Geol. Soc., Cleveland, Ohio, 181-200.

Jones, B.F., Eugster, H.P., and Rettig, S.L. (1977) Hydrochemistry of the Lake Magadi Basin, Kenya: Geochim. Cosmochim. Acta 41, 53-72.

Kossovskaya, A.G. (1975) Genetic types of zeolites in stratified rocks: Lithology and Mineral. Res. 10, 23-44.

Langbein, W.B. (1961) Salinity and hydrology of closed lakes: U. S. Geol. Surv. Prof. Paper 412, 20 pp.

Mariner, R.H. (1971) Experimental evaluation of authigenic mineral reactions in the Pliocene Moonstone Formation: Ph.D. thesis, University of Wyoming. 133 pp.

Mariner, R.H., and Surdam, R.C. (1970) Alkalinity and formation of zeolites in saline alkaline lakes: Science 170, 977-980.

Maxey, G.B. (1968) Hydrology of desert basins: Ground Water 6, 10-22.

Mumpton, F.A. (1973a) Scanning electron microscopy and the origin of sedimentary zeolites: In, Molecular Sieves, Proc. 3rd Int. Conf. Mol. Sieves, Uytterhoeven, Ed., Leuven University Press, 56-61.

Mumpton, F.A. (1973b) Worldwide deposits and utilisation of natural zeolites: Industrial Minerals 73, 30-45.

Mumpton, F.A., and Ormsby, W.C. (1976) Morphology of zeolites in sedimentary rocks by scanning electron microscopy: Clays and Clay Minerals 24, 1-23.

Munson, R.A., and Sheppard, R.A. (1974) Natural zeolites: their properties, occurrences, and uses: Mineral Sci. Eng. 6, 19-34.

Ross, C.S. (1928) Sedimentary analcite: Am. Mineral. 13, 195-197.

Russell, I.C. (1896) Present and extinct lakes of Nevada: In, The Physiography of the United States, American Book Co., New York, 288 pp.

Senderov, E.E. (1963) Crystallization of mordenite under hydrothermal conditions: Geochem. 9, 848-859.

Sheppard, R.A. (1971) Zeolites in sedimentary deposits of the United States--a review: In, Gould, R.F., Ed., Molecular Sieve Zeolites--1: Am. Chem. Soc., Advances in Chemistry Series 101, 279-310.

Sheppard, R.A. (1973) Zeolites in sedimentary rocks: U. S. Geol. Surv. Prof. Paper 820, 689-695.

Sheppard, R.A. (1975) Zeolites in sedimentary rocks: In, Leford, S.J., ed., Industrial Minerals and Rocks, 4th ed., Am. Inst. Mining, Metal. and Petroleum Engineers, Inc., New York, 1257-1262.

Sheppard, R.A., and Gude, A.J., 3d (1968) Distribution and genesis of authigenic silicate minerals in tuffs of Pleistocene Lake Tecopa, Inyo County, California: U. S. Geol. Surv. Prof. Paper 597, 38 pp.

Sheppard, R.A., and Gude, A.J., 3d (1969) Diagenesis of tuffs in the Barstow Formation, Mud Hills, San Bernardino County, California: U. S. Geol. Surv. Prof. Paper 634, 35 pp.

Sheppard, R.A., and Gude, A.J., 3d (1973) Zeolites and associated authigenic silicate minerals in tuffaceous rocks of the Big Sandy Formation, Mohave County, Arizona: U. S. Geol. Surv. Prof. Paper 830, 36 pp.

Smith, C.L., and Drever, J.I. (1976) Controls on the chemistry of springs at Teels Marsh, Mineral County, Nevada: Geochim. Cosmochim. Acta 40, 1081-1093.

Smith, G.I. (1966) Geology of Searles Lake: In, Rau, J.L., Ed., Second Symposium on Salt, Northern Ohio Geol. Soc. 1, 167-180.

Stumm, W., and Morgan, J.J. (1970) Aquatic Chemistry: John Wiley, New York, 583 pp.

Surdam, R.C., and Eugster, H.P. (1976) Mineral reactions in the sedimentary deposits of the Lake Magadi region, Kenya: Geol. Soc. Am. Bull. 87, 1739-1752.

Surdam, R.C., and Mariner, R.H. (1971) The genesis of phillipsite in Recent tuffs at Teels Marsh, Nevada: Geol. Soc. Am. Abstr. Progr. 3.

Surdam, R.C., and Parker, R.B. (1972) Authigenic aluminosilicate minerals in the tuffaceous rocks of the Green River Formation, Wyoming: Geol. Soc. Am. Bull. 83, 689-700.

Surdam, R.C., and Sheppard, R.A. (1978) Zeolites in saline, alkaline lake deposits: In, Sand, L.B. and Mumpton, F.A., Eds., Natural Zeolites: Occurrence, Properties, Use, Pergamon Press, Elmsford, New York, 145-174.

91

Chapter 5

ZEOLITES IN OPEN HYDROLOGIC SYSTEMS

R. L. Hay and R. A. Sheppard

INTRODUCTION

In contrast to those deposits formed by the alteration of volcanic ash in saline, alkaline lakes, large masses of tuffaceous sediments have been transformed to zeolite minerals by the action of percolating water in open hydrologic systems. Zeolite deposits of the open-system type are commonly several hundred meters thick and can be traced laterally for several tens of kilometers. Thick zones containing up to 90% clinoptilolite or mordenite are not uncommon in some bodies; hence, open-system deposits have considerable economic significance. Open-system deposits are most abundant in nonmarine rocks; however, many occurrences are known in sediments which were deposited in shallow marine environments. The original pyroclastic material was laid down in the sea fairly close to the volcanic source or air-laid onto the land surface. Several examples of the latter type have been found in the United States, but most of the large marine bodies occur in Japan and in southern and southeastern Europe. Due to their areal extent, open-system deposits have not been studied in as much detail as many other types of zeolite deposits in sedimentary rocks; however, studies of several key areas have provided a sound basis for our current understanding of this type of zeolite body.

Zeolitic alteration can take place in tephra deposits when flowing or percolating ground water becomes chemically modified by hydrolysis or dissolution of vitric materials. The formation of clay minerals or palagonite results in the release of hydroxyl ions to the ground water, and the solution becomes increasingly alkaline and enriched in Na, K, and Si along its flow path. Zeolite minerals are formed in addition to or instead of clay minerals where the cation to hydrogen ion ratio and other ionic activities are appropriate. Meteoric water entering the system moves either downward or with a downward component; hence, the zeolitic alteration zones are either horizontal or gently inclined. Zeolitic alteration zones in open-system deposits commonly cut across stratigraphic boundaries and show more-or-less vertical sequences of authigenic silicate minerals.

93

This zoning is analagous to the leaching experiment of Morey and Fournier (1961) in which distilled water was pumped slowly through a column of finely ground nepheline at 295°C. When the experiment was terminated, boehmite and paragonite had replaced the nepheline at the inlet end of the column, and muscovite and analcime had formed at the exit end where the water had attained a pH of 9.7 and 440 ppm of dissolved solids. Using rhyolitic glass as the starting material, Wirsching (1976) simulated open-system alteration. Experiments with 1 g of glass in 50 ml of 0.01 N NaOH at 150°C showed that minor phillipsite and a trace of mordenite formed after 32 days. With increasing time, phillipsite and mordenite increased in abundance, and a trace of analcime was detected after 42 days. At the end of the 80-day experiment, analcime was the dominant phase. At higher temperatures, analcime and alkali feldspar were the final products. The pH of the experiments was about 10.

In natural ground waters, where the pH is raised through hydrolysis, the reaction rate of volcanic glass is vastly increased. In saline, alkaline lakes, where the pH is 9.5, tuffs can be wholly altered in 10^3 to 10^4 years. Zeolite zones produced at high pH characteristically have relatively sharp upper contacts with rocks containing fresh glass; whereas, zeolitic zoning produced at lower pH has gradational contacts. The zones may be further zoned mineralogically as a result of the reaction of early-formed zeolites with the charged ground water in the lower part of the tuff sequence. Clinoptilolite, mordenite, and minor phillipsite are among the most common of the first-formed zeolites in open systems, and analcime and/or potassium feldspar characterize the deeper zones. The exact zeolite assemblage and the composition of individual zeolite species are controlled, at least in part, by the composition of the precursor tephra. Erionite is generally absent from zeolite deposits formed in open hydrologic systems.

ZEOLITE FORMATION IN SILICIC TEPHRA DEPOSITS

Although numerous deposits of zeolitic tuff are known throughout the western United States and in other countries which must have been formed in open hydrologic basins, nonmarine accumulations of silicic tephra must reach thicknesses of a few hundred meters to develop widespread zeolitic zoning as described above. The John Day Formation (Oligocene-Miocene) of central Oregon is an excellent example of this type and illustrates well the sequence and conditions of zeolite formation and the processes by which silicic glass is altered to clinoptilolite. The John Day Formation consists of 600-900 m of land-laid silicic tuff and tuffaceous claystones (Hay, 1963).

Zeolites have been identified in the lower part of this formation in an area of about 5,700 km^2. Where the top of the formation is eroded least, an upper zone, 300-450 m thick, contains either fresh glass or glass altered to montmorillonite and opal (see Figure 1). Clinoptilolite with lesser amounts of montmorillonite, celadonite, and opal replaces nearly all of the glass in the lower 300-450 m. Potassium feldspar is locally common in the lower part of the zeolite zone, chiefly as a replacement of pyrogenic plagioclase.

The upper surface of the zeolitic alteration is nearly horizontal and cuts across stratification where the beds are folded. The transition between the upper and lower zones is relatively sharp and occurs through a cavernous zone 2-20 mm thick where glass shards have been dissolved. Zeolite replacement of shards took place by the crystallization of laths and platelets of clinoptilolite, ranging from 0.01-0.1 mm in size, in pseudomorphic cavities from which glass was dissolving or had already been dissolved. The transition between the zeolite zone and the fresh glass zone may follow a single stratigraphic horizon over distances of as much as 5 km, but elsewhere it shows an interfingering relationship.

Spheroids or concretions of zeolite tuff near the base of the upper zone provide clear evidence that hydrolysis of glass to montmorillonite can help create the chemical

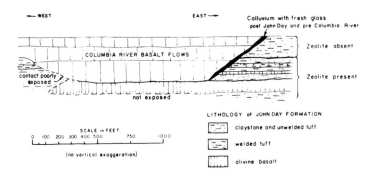

Figure 1. Lower part of a valley cut in beds of the John Day Formation and filled with Columbia River basalt flows. Colluvium mantling the valley wall at the right is a clayey tuff containing fresh glass and lacking zeolite. It lies adjacent to zeolitic claystones of the John Day Formation, thus proving that zeolitic alteration preceded the colluvium, as well as the Columbia River Basalt flows. Reprinted from Hay (1963) by permission of the University of California Press.

environment in which glass reacts to form zeolites. The concretions are 2-10 cm in diameter and lie in a matrix of fresh vitric ash of an unwelded ignimbrite. A large fragment of pumice altered to montmorillonite forms the nucleus of each concretion. These relationships suggest that the alteration of pumice to montmorillonite locally raised the pH and ionic activities to a level where the adjacent glass reacted, ultimately forming clinoptilolite.

The replacement of rhyolitic glass by clinoptilolite involved a gain of H_2O, Ca, and Mg and a loss of Si, Na, K, and Fe. Altered vitric rocks of original rhyolitic composition in the John Day Formation gained Fe, Ti, Mg, Ca, and less commonly, P. Some of the alkalies were lost; silica was lost from about half of the altered rhyolitic rocks, but remained nearly constant in the others.

K-Ar dating suggests that the zeolites and associated minerals were formed only after most of the sediments had been deposited, and zeolitic alteration may have lasted for a relatively short time. Dates from primary eruptive materials range from 32 m.y.a. at the base to 25 m.y.a. slightly above the middle of the formation. The upper part of the formation may well be as young as 20 m.y.a. on the basis of faunal content. Dates of 22 and 24 m.y.a. were given by authigenic K-feldspar and celadonite from the zeolitic zone suggesting that alteration took place only after about 600 m of tuffs had been deposited. Zeolitic alteration had probably ceased before the time of deepest burial, about 15 m.y.a. when the Columbia River Basalt was erupted (Figure 1).

Volcanic sedimentary rocks of the Vieja Group (lower Oligocene) of Trans-Pecos Texas are about 800 m thick and show a zonation of silicate minerals that probably resulted from diagenetic alteration of the open-system type (Walton, 1975). Clinoptilolite, montmorillonite, opal, quartz, and analcime formed in silicic volcaniclastic sediments from constituents derived from the volcanic glass. From top to bottom, the vertical zonation is characterized by the presence of (1) fresh glass, (2) clinoptilolite, and (3) analcime (see Figure 2). Montmorillonite and opal are locally present in the fresh glass zone, montmorillonite and opal or quartz in the clinoptilolite zone, and montmorillonite and/or quartz in the analcime zone.

The fresh glass zone has a maximum thickness of 220 m. Here, thin films of montmorillonite or opal coat hydrated glass shards. Such coatings persist in the clinoptilolite zone and appear to be responsible for the preservation of the shape of the shards during replacement by zeolite and montmorillonite. In the clinoptilolite zone, clay or clinoptilolite fill much of the remaining intergranular space. In the analcime zone, clinoptilolite is replaced by analcime in both shards and cement. A large amount of aluminum must have been added during the conversion of clinoptilolite to analcime, presumably from the dissolution of montmorillonite. The replacement of

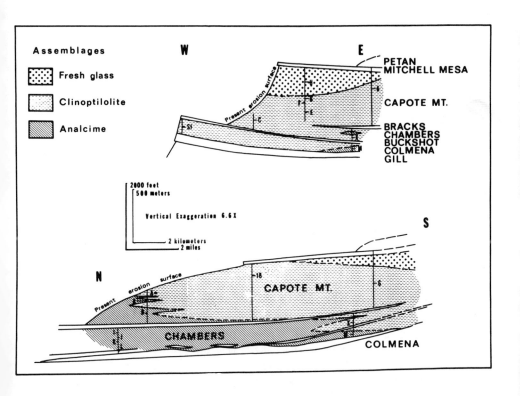

Figure 2. Schematic cross section of Vieja Group in Candelaria area of southern Rim Rock Country, Texas, drawn to remove effects of Miocene and later faulting. Distribution of diagenetic assemblages was determined by x-ray and thin section studies of samples from lettered sections and isolated samples between. Dashed lines extrapolate boundaries beyond areas of intensive control. Reprinted from Walton (1975) by permission of the Geological Society of America.

clinoptilolite by analcime took place at constant volume as evidenced by the similarity of textures of rocks in both zones. In contrast to the John Day Formation, potassium feldspar is essentially absent from the Vieja Group sediments. All authigenic silicates formed at burial depths no greater than a few hundred meters. In the northeastern part of the area, the rocks were less deeply buried than in the southeastern part of the region, and the boundaries of the alteration zones are stratigraphically higher. The zeolites may have formed while the deposits were accumulating, and the alteration zones were offset by early Miocene faulting. Boundaries between diagenetic zones may also reflect permeability differences among the rocks.

The authigenic mineral zonation in Tertiary tuffs at the Nevada Test Site (Hoover, 1968) is more complex than either that of the John Day Formation or the Vieja Group. Altered rocks in which clinoptilolite or analcime predominate are present throughout an area of several thousand square kilometers and range in thickness from a few tens of meters to about 2000 m. Incomplete zonation of silicate minerals cuts across stratigraphic boundaries but corresponds in a general way to the present water levels in the region. An upper zone is characterized by the presence of fresh glass and contains local concentrations of clay minerals or chabazite. It overlies a succession of zones rich in (1) clinoptilolite, (2) mordenite, and (3) analcime, in descending order (see Figure 3). Cristobalite is generally restricted to the upper part of the clinoptilolite zone and higher zones. Authigenic quartz and potassium feldspar are relatively abundant in the lower part of the clinoptilolite zone and in the analcime zone. Temperatures measured in drill holes through the zeolitic rocks range from 25°-65°C, and there is no evidence for higher temperatures during zeolitization. A hydrothermal or burial metamorphic origin is considered unlikely, and a leaching-deposition mechanism of the open hydrologic system type is postulated for this body, possibly controlled by ground water levels and permeability barriers in the section.

Other probable examples of open-system zeolite bodies are common in tuffaceous sequences of Miocene and Pliocene age in the western United States. The Ricardo Formation (Pliocene) in Kern County, California, is an example of open-system type zoning in these younger sediments. Here, the beds have been tilted, yet the contact between fresh and zeolitic tuff is nearly horizontal, showing that alteration occurred after tilting (Sheppard and Gude, 1965). Clinoptilolite is the only zeolite in this sequence, and it is associated with variable amounts of montmorillonite and opal.

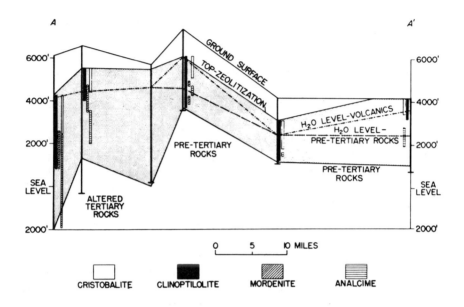

Figure 3. Cross section of tuffaceous rocks of the Nevada Test Site showing the
relationship of diagenetic alteration zones. (From Hoover, 1968.)

ZEOLITE FORMATION IN LOW-SILICA TEPHRA DEPOSITS

Alkaline, low-silica tuffs only 10-20 m thick may be zoned zeolitically to the
same extent as silicic tuffs 500-1,000 m thick. Zonation at shallow depth reflects
the highly reactive nature of alkalic, low-silica glass. The formation of zeolites
from mafic glasses in open hydrologic systems is somewhat more complex than it is
from silicic glasses. Mafic glass (or sideromelane) reacts to form palagonite and
zeolites in an appropriate chemical environment which is generally alkaline. Palagonite can be viewed as a hydrous, iron-rich gel formed from mafic volcanic glass.
It contains somewhat less Al, Si, Ca, Na, and K than the original glass. These components provide the raw materials for the crystallization of zeolites in pore spaces.
Phillipsite and chabazite are probably the most common zeolites in low-silica tuffs,
but natrolite, gonnardite, analcime, and other low-silica zeolites are not uncommon.

Koko Crater, Hawaii, illustrates the alteration of alkaline, mafic tephra in an open-system environment. Koko Crater is one of a group of craters formed on southern Oahu about 35,000 years ago. Its cone consists largely of tuffs and lapilli tuffs of basanite composition.[1] An upper zone characterized by the presence of fresh glass, opal, and montmorillonite overlies a lower zone of zeolitic, palagonitic tuff (Hay and Iijima, 1968). The upper zone is generally 20-12 m thick, and the transition between the two zones is a few centimeters to a meter thick. Two zeolitic sub-zones have been recognized within the palagonite-zeolite zone. The upper sub-zone contains phillipsite and chabazite, and the lower sub-zone contains phillipsite and analcime. A striking feature of the palagonitic alteration is the loss of about half of the aluminum from the mafic glass and its precipitation in the form of interstitial zeolites.

The contact between zones roughly parallels the topography and cuts across stratification (Figure 4), indicating that the palagonitic alteration is a post-eruptive phenomenon, related to percolating meteoric water rather than to hydrothermal solutions. Palagonitic tuff abruptly grades downward into relatively fresh tuff approximately at sea level, presumably reflecting dilution of reactive alkaline meteoric water with sea water of lower pH. Under these conditions, the glass reacted more slowly. As it

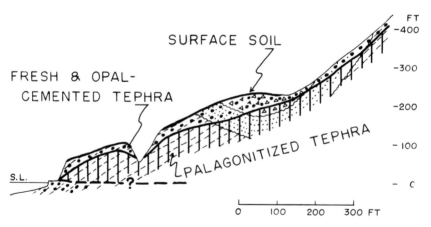

Figure 4. Cross section through the southeastern edge of the cone of Koko Crater, showing palagonitic alteration which transects stratification and ends abruptly at about sea level (abbreviated SL). Reprinted from Hay and Iijima (1968) by permission of the Geological Society of America.

----

[1]Basanite is a mafic igneous rock whose composition is close to that of basalt but which contains normative nepheline.

descended through the column, the meteoric water may have reached a pH as high as 9.5, which could account for the extensive degree of alteration in less than 35,000 years and for the relatively sharp transition between fresh and palagonitic tuff. A high pH could also explain the mobility of Al, which is relatively soluble in the form of $Al(OH)_4^-$ under these conditions.

The alteration in the Salt Lake Tuff, also on Oahu, is related to the original thickness of tuff and to the position of the tuffs relative to sea level in the Pleistocene. Where the nephelinite tuffs thin to the north and west, zeolitic pala- gonite intergrades laterally with fresh tuff. Fresh tuff commonly overlies palagonite tuff in the transition zone (Figure 5). Where the tuffs are thinner than 9 m, they are unaltered except for surface weathering; where the tuffs are thicker than 9 m, the upper 4-5 m are fresh, and the underlying material is palagonitized and cemented by zeolites. The vertical transition between fresh and palagonite tuff is generally sharp. The principal zeolites in the Salt Lake Tuff are phillipsite, natrolite, gonnardite, and analcime. The Salt Lake Tuff was deposited when sea level was lower than at present, and the tuff was palagonitized and cemented by zeolites before the sea rose in late Pleistocene time from 6-8 m above the present level.

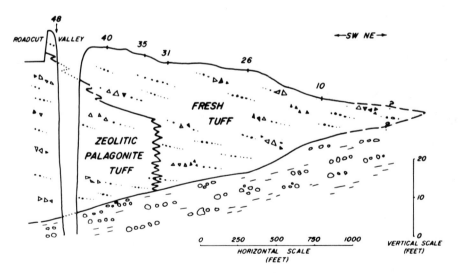

Figure 5.   Salt Lake Tuff exposed along the northwest side of Moanalua Stream, at the northeast corner of the Salt Lake Tuff. Numbers along the surface of the Salt Lake Tuff represent thicknesses of measured sections. Boulder con- glomerates, claystones, and reworked nephelinite ash layers underlie the Salt Lake Tuff. Reprinted from Hay and Iijima (1968) by permission of the Geological Society of America.

The Neapolitan Yellow Tuff (tuffo giallo napoletano) near Naples, Italy, provides an example of the formation of zeolites in an open-system environment from tuff of trachytic composition (see Sersale, 1958). This formation, which underlies the city of Naples, crops out over an area of about 200 km^2, and locally reaches more than 100 m in thickness. In the lower part of the thickest exposures, glass is completely altered to phillipsite and chabazite. The overlying material is essentially zeolite free and is locally known as "pozzolana." The transition from glass to zeolite is a continuous one, with the zeolite-rich tuff being characterized by a higher water content and a lower alkali content. The tuff erupted about 10,000 to 12,000 years ago, and isotopic evidence suggests that zeolitic alteration lasted from 4000 to 5000 years. It is not now thought to be active (Capaldi et al., 1971).

REFERENCES

Capaldi, G., Civetta, L., and Gasparini, P. (1971) Fractionation of the [238]U decay series in the zeolitization of volcanic ashes: Geochim. Cosmochim. Acta 35, 1067-1072.

Hay, R.L. (1963) Stratigraphy and zeolitic diagenesis of the John Day Formation of Oregon: Calif. Univ. Publs. Geol. Sci. 42, 199-262.

Hay, R.L., and Iijima, A. (1968) Nature and origin of palagonite tuffs of the Honolulu Group on Oahu, Hawaii: Geol. Soc. Am. Mem. 116, 331-376.

Hoover, D.L. (1968) Genesis of zeolites, Nevada Test Site: Geol. Soc. Am. Mem. 110, 275-284.

Morey, G.W., and Fournier, R.O. (1961) The decomposition of microcline, albite, and nepheline in hot water: Am. Mineral. 46, 688-699.

Sheppard, R.A., and Gude, A.J., 3d (1965) Zeolitic authigenesis of tuffs in the Ricardo Formation, Kern County, southern California: U. S. Geol. Surv. Prof. Paper 525, D44-D47.

Sersale, R. (1958) Genesi et costituzione del tufo giallo napoletano: Rend. Accad. Sci. Fis. Mat. 25, 181-207.

Walton, A.W. (1975) Zeolitic diagenesis in Oligocene volcanic sediments, Trans-Pecos Texas: Geol. Soc. Am. Bull. 86, 615-624.

Wirsching, U. (1976) Experiments on hydrothermal processes of rhyolitic glass in closed and "open" system: Neues Jahrb. Mineral. Monatsch. 5, 203-213.

Chapter 6

ZEOLITES IN LOW-GRADE METAMORPHIC ROCKS

James R. Boles

INTRODUCTION

Studies of zeolites in low-grade metamorphic rocks are important to an understanding of the transition between diagenesis and conventional metamorphism. Zeolite assemblages give clues to the chemistry and mobility of pore fluids and the temperature history that the rocks have undergone in this transition zone.

Zeolites in low-grade metamorphic rocks occur in two types of terraines: (1) hydrothermal and (2) burial. Hydrothermal occurrences include active and fossil geothermal systems and rocks hydrothermally altered by igneous intrusion. Zeolites developed on a regional scale in thick stratigraphic sections are usually attributed to burial metamorphism. In many of these areas igneous intrusions are common, and, in such cases, it is difficult to distinguish burial from hydrothermal effects. The emphasis here will be on burial metamorphic occurrences of zeolites, but examples in active geothermal areas will also be discussed because, unlike most burial settings, alteration conditions are reasonably well known.

The regional occurrence of zeolites in low-grade metamorphic rocks was first recognized by Coombs (1954) in southern New Zealand. Largely on the basis of these observations, Turner (in Fyfe et al., 1958) defined a zeolite facies to include regionally developed assemblages of laumontite-albite-quartz. The zeolite facies was believed to represent P-T conditions intermediate between diagenesis and conventional metamorphism. Coombs et al. (1959) treated zeolite assemblages in terms of Eskola's concept of mineral facies (1920) which allowed the facies to embrace products of diagenesis, hydrothermal activity, or metamorphism. They broadened the definition to include assemblages of analcime-quartz, heulandite-quartz, and laumontite-quartz. More recently, Coombs (1971, p. 324) suggested that the zeolite facies might be defined as ". . . that set of mineral assemblages that is characterized by the association calcium zeolite-chlorite-quartz in rocks of favorable bulk composition."

The zeolite facies is a useful concept in that it acknowledges intermediate stages between diagenesis and conventional metamorphism. However, at present we only crudely understand the pressure, temperature, and chemical significance of various zeolite mineral assemblages in such rocks. Clearly the presence of some zeolites (e.g., laumontite and wairakite) represents more advanced alteration than others (e.g., heulandite), but physiochemical conditions at which the alteration occurred are less obvious In many zeolite-bearing metamorphic terraines it is likely that diagenetic and low-grade metamorphic assemblages coexist.

Several generalizations can be made about zeolite occurrences in low-grade metamorphic terraines:

(1) Zeolites most commonly occur as alteration products of volcanic glass. Calcic plagioclase is also an important precursor. These types of detrital clasts are particularly common in, but not necessarily restricted to, Mesozoic and Tertiary rocks of the circum-Pacific region.

(2) Zeolites commonly make up less than 25% of the bulk rock, except in altered vitric tuffs where much higher percentages may be found. The zeolite-bearing rocks retain much of their detrital mineralogy and texture and are rarely completely recrystallized, in contrast to the rocks in the greenschist facies. Schistosity and penetrating cleavage are absent.

(3) The distribution of individual zeolites with respect to depth of burial commonly shows considerable overlap, which clearly indicates that factors other than temperature and pressure are also important reaction controls.

(4) On a worldwide basis, laumontite, analcime, and heulandite-group[1] minerals are the most common zeolites in low-grade metamorphic rocks. Tertiary and younger rocks may contain more diverse zeolitic assemblages, including the minerals mordenite, erionite, chabazite, and phillipsite.

(5) The zeolites are generally associated with authigenic quartz, albite, adularia, calcite, phyllosilicates and less commonly with sphene, prehnite, and pumpellyite.

Although low-grade metamorphic zeolite assemblages have been reported in a number of other areas, including New Brunswick (Mossman and Bachinski, 1972), Puerto Rico

----

[1]Heulandite group will be used in this review to include both clinoptilolite and heulandite.

(Otálaró, 1964; Jolly, 1970), France (Martini and Vaugnat, 1968), and Russia (Zaporozhtseva, 1960; Zaporozhtseva et al., 1961; Kossovskaya and Shutov, 1961), most reported
occurrences are in the circum-Pacific area.

SELECTED ZEOLITE OCCURRENCES FROM THE CIRCUM-PACIFIC AREA

New Zealand

The north limb of the Southland Syncline is composed of about 9-10 km of Triassic
and Jurassic marine strata (Figure 1). The thickness of the section, its structural
simplicity, and the lack of in situ igneous rocks in the strata make this area particularly suitable for burial metamorphic studies. Vitric and vitric-crystal tuffs occur
through the section. Volcanic glass, once common in the sequence, has been completely
replaced. Sandstones in the area are highly volcanogenic and/or feldspathic (chiefly
plagioclase) and generally have less than 20% quartz relative to feldspar and rock
fragments.

Coombs (1954) presented the first comprehensive description of alteration of the
Triassic rocks in the Taringatura area, summarized as follows by Coombs et al. (1959,
p. 59):

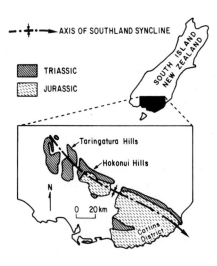

Figure 1. Index map of Mesozoic
rocks in the Southland Syncline,
New Zealand.

"Coombs (1954) showed that glass in ash beds in the upper part of the
section has been completely replaced by heulandite or less commonly by analcime. Both zeolites coexist with quartz and fine-grained phyllosilicates.
The great majority of rocks here contain fresh detrital lime-bearing plagioclase as a major constituent, but from successively lower horizons it is
found to be missing from more and more specimens. In its place are pseudomorphs of dusty albite with sericite inclusions, although some 'islands'
rich in relict andesine persist almost to the base of the section. Simultaneously with the albitization of plagioclase, analcime and heulandite
disappear. Analcime has not been observed in rocks from lower than about
17,000 ft. below the top of the section, its place being taken mainly by
albite, but also in some cases by pseudomorphs of adularia or laumontite.
Heulandite persists to greater depths, but it appears to give way to laumontite plus quartz, and the lower part of the section contains numerous
beds of altered ash, some of them very thick, in which laumontite is the
dominant constituent. First pumpellyite, then prehnite appear as accessory
minerals. In some cases at least, notably in quartz-albite-adularia-pum-
pellyite metasomatites, the pumpellyite has clearly formed at the expense
of laumontite, magnesium and iron being provided by celadonite and chloritic
minerals."

In addition to laumontite pseudomorphing glass shards, Coombs (1954, p. 74) recog-

nized that ". . . laumontite is a joint replacement product with albite of lime-bearing

plagioclase. Directly or indirectly, the breakdown of lime-bearing plagioclase is a

main feature in the origin of most laumontite." Coombs (1954) attributed the altera-

tion sequence chiefly to increasing temperature associated with progressive burial.

The Hokonui Hills area (Figure 1) includes more extensive exposures of Jurassic

and uppermost Triassic strata than can be observed in the Taringatura area. Recent

work in the former area by Boles (1971, 1974) and Boles and Coombs (1975, 1977) has

shown that laumontite and heulandite are common in both the lower and uppermost parts

of the section. In sandstones, authigenic heulandite, laumontite, prehnite, sphene,

chlorite, and pumpellyite are associated with rocks of andesitic parentage, whereas

authigenic quartz, albite, and K-feldspar characterize sandstones of rhyolitic to

dacitic parentage. Mineral distribution patterns in the two areas are summarized in

Figure 2. The amount of overburden eroded from this section is uncertain but is be-

lieved to be less than 1.5 km. Distribution patterns and stability relations of anal-

cime with quartz and of laumontite show that average temperature gradients did not

exceed about 25°C/km. Boles and Coombs (1977) concluded that the complexity of mineral

distribution patterns is due to the interplay of many factors, including the nature of

parent material including glass and unstable minerals, incomplete reactions, perme-

ability, ionic activity in pore fluids, P_{CO_2}, varying $P_{fluid}P/_{total}$ ratios, and effects

of rising temperature following burial.

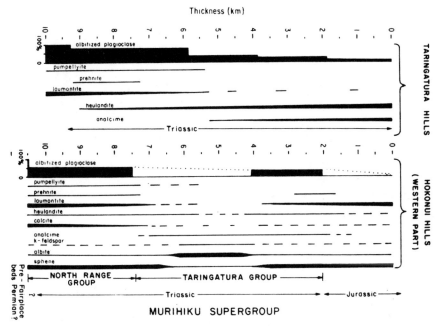

Figure 2. Mineral distribution patterns in the Taringatura and Hokonui Hills, New
Zealand (after Coombs, 1954; Boles and Coombs, 1977). The "albitized plagio-
clase" of the Hokonui Hills represents percentage of sandstones (0.2-0.5 mm
average grain size) with essentially all plagioclase grains albitized.
Hokonui Hills "albite" refers to authigenic albite in cavities and as cement.
Dashed line denotes sporadic occurrence.

Japan

Zeolites have been reported in sediments ranging from Quaternary to Triassic age in
Japan, but the majority of the occurrences are in Miocene pyroclastic and volcanogenic
sedimentary rocks (Figure 3). The zeolitic rocks are usually of rhyolitic to dacitic
parentage and are largely of marine origin. Japanese zeolite occurrences have been extensively studied by many workers and recently reviewed by Utada (1970, 1971) and Iijima
and Utada (1972).

Zeolite occurrences in Japan are largely attributed to low-grade metamorphism, and
distinct zeolite zonations are commonly recognized. Some of these zonations are attributed to burial diagenesis; others are attributed to hydrothermal effects of igneous
intrusions; and still others are attributed to modern and fossil geothermal waters.
Many occurrences are probably a result of several of these factors.

Figure 3. Index map of selected zeolite occurrences in low-grade metamorphic rocks of Japan. See text for description of localities.

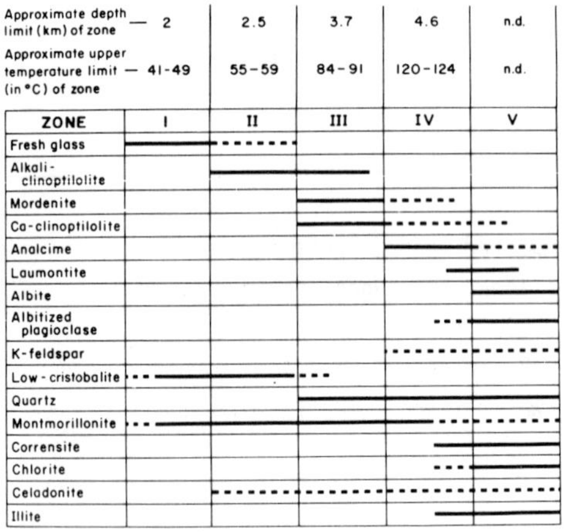

Niigata oil field

Katayama geothermal area

JAPAN

TOKYO

Tanzawa Mountain area

Motojuku district

0   200 km

	I	II	III	IV	V
Approximate depth limit (km) of zone — 2	2.5	3.7	4.6	n.d.	
Approximate upper temperature limit (in °C) of zone — 41-49	55-59	84-91	120-124	n.d.	
ZONE	I	II	III	IV	V
Fresh glass					
Alkali-clinoptilolite					
Mordenite					
Ca-clinoptilolite					
Analcime					
Laumontite					
Albite					
Albitized plagioclase					
K-feldspar					
Low-cristobalite					
Quartz					
Montmorillonite					
Corrensite					
Chlorite					
Celadonite					
Illite					

Figure 4. Stability ranges of authigenic minerals in the Niigata oil field (modified after Iijima and Utada, 1972). Dashed line denotes rare occurrence.

108

In burial metamorphic terraines, five depth-related zones with characteristic phases are found: Zone I--fresh glass; Zone II--alkali clinoptilolite; Zone III--clinoptilolite + mordenite; Zone IV--analcime ($\pm$ heulandite); and Zone V--albite. In addition, laumontite can occur either between or overlapping Zones IV and V. The distribution of authigenic minerals and the temperature-depth relations of the zones in the Niigata oil field are shown in Figure 4. The characteristic authigenic minerals of the zones occur chiefly as replacements of volcanic glass, although plagioclase replacement occurs in Zones IV and V.

Several notable aspects of the above zonations which differ from most other low-grade zeolite terraines in the circum-Pacific area are:

(1) Mordenite is relatively common and widespread in low-grade metamorphic rocks of Japan.

(2) Opaline silica (cristobalite) commonly coexists with clinoptilolite and mordenite (Utada, 1970).

(3) A transition from alkalic clinoptilolite to Ca-clinoptilolite or heulandite is inferred with increasing burial depth in some areas (see Figure 4).

(4) Laumontite is not particularly common in these rocks even in the higher grade zones (cf. Minato and Utada, 1969, p. 126). Vitric tuffs are commonly replaced by albite + quartz assemblages in the higher grade rocks rather than by calcium-aluminosilicates (Utada, 1970, p. 230).

Miocene and younger pyroclastic sedimentary rocks of Japan intruded by granitic or dioritic rocks, for example in the Tanzawa Mountains and in the Motojuku district of North Kanto (Figure 3), also show zeolite alteration zones in the surrounding sediments. Mineral zonations in the Tanzawa area are shown in Figure 5. Compared with the burial diagenesis occurrences described above, the zones have a greater variety of zeolites, many of which are calcium rich. Wairakite is a common zeolite in the higher grade zeolite-bearing zones.

In the Tanzawa Mountains, the zones are more or less concentric to the intrusive body and regional cross bedding (see Figure 5a). The lower grade limit of Zone II in the Tanzawa Mountains ranges from about 6 to 9 km from the intrusive body. Thus, the thermal effects of contact metamorphism may cause zeolitization (in this case laumontite) at a considerable distance from the intrusive body.

Zeolite zonations have also been recognized in active geothermal areas of Japan. For example, Seki et al. (1969b) described three zeolite alteration zones in interbedded andesitic lavas and dacitic tuffs from the Katayama geothermal area. The shallowest

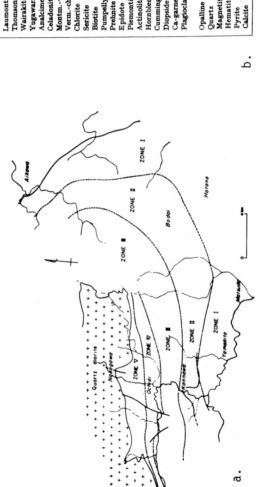

ZONE	I	II	III	IV	V
Clinoptilolite					
Stilbite		>>	>>	>	>
Heulandite					
Mordenite					
Chabazite					
Laumontite		>	>>		
Thomsonite					
Wairakite					
Yugawaralite					
Analcime					
Celadonite					
Montm.-verm.					
Verm.-chlorite					
Chlorite					
Sericite					
Biotite					
Pumpellyite					
Prehnite					
Epidote					
Piemontite					
Actinolite					
Hornblende					
Cummingtonite					
Diopside					
Ca-garnet					
Plagioclase				An 10 20 30	
Opalline silica					
Quartz					
Magnetite					
Hematite					
Pyrite					
Calcite					

b.

a.

Figure 5. (a) Distribution of mineral zones about a quartz diorite intrusion, Tanzawa Mountains, Japan
(from Seki et al., 1969a).
(b) Mineral zones in Tanzawa Mountains, Japan (from Seki et al., 1969a). "V" denotes minerals
found in veins. Dashed line denotes not common.

zone is characterized by mordenite, the next by laumontite (in places accompanied by analcime or yugawaralite), and the third by wairakite. The temperature limits of these occurrences vary from drill hole to drill hole (see Figure 6). Notably, wairakite first occurs at temperatures ranging from about 75°-175°C; this is in contrast to its occurrence in geothermal wells of Wairakei, New Zealand, where it does not appear to have formed below about 200°C (see Coombs et al., 1959, p. 71).

From the above discussions one may conclude that zeolite occurrences in low-grade metamorphic rocks of Japan developed largely in relatively high T/P regimes during late Tertiary time. The high thermal gradients may have been responsible for the distinct zeolite zonations that are found in Japan but which are less common in other areas of the world.

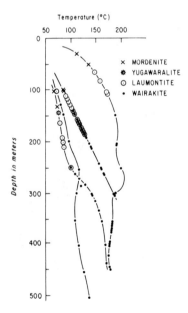

Figure 6. Geothermal gradients and zeolite distributions in the Katayama geothermal area, Onikobe, Japan. (From Seki et al., 1969b.)

Western North America

British Columbia - Triassic rocks. Surdam (1973) described zeolitization in the Karmutsen Group, a 5.5-km thick marine sequence of pillow lava and breccia, aquagene tuff, massive amygdaloidal flows, and interlava limestones (Figure 7). An estimated 1.5 to 4.6 km of overburden has been removed from the top of the section. Surdam recognized increased albitization of plagioclase and the breakdown of analcime to albite with increasing depth in the section. Laumontite and wairakite replace calcic plagioclase in the matrix of once glassy rocks and in amygdales of volcanic flows. Clays rather than zeolites form pseudomorphs after volcanic glass.

Surdam (1973) attributed much of the detailed overlap of mineral assemblages, such as closely spaced occurrences of laumontite and wairakite, to minor variations in activities of ionic species in solution. Judging from the abundant in situ igneous rocks and the presence of wairakite in the Karmutsen section, the geothermal gradient in the area was probably relatively high.

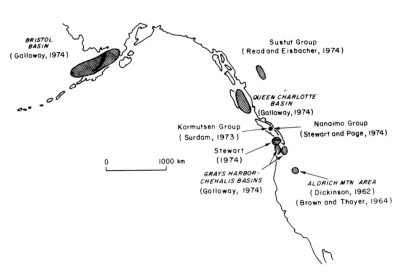

Figure 7. Index map of selected zeolite occurrences in low-grade metamorphic rocks of western North America. See text for description of localities.

British Columbia - Cretaceous rocks. Stewart and Page (1974) described zeolitization of sandstones in the Upper Cretaceous Nanaimo Group, an interbedded sequence of
lagoonal to deltaic to marine sediments up to 3 km thick (Figure 7). Minor igneous
rocks of Tertiary age intrude the group. The Nanaimo Group overlies the Karmutsen
Group described by Surdam (1973).

The sandstones typically contain framework grains of 40% quartz, 50% feldspar
(mainly plagioclase), and 10% aphanitic fragments. Phyllosilicates, heulandite, laumontite, and carbonates make up the matrix of the sandstones. Laumontite also occurs
as a replacement of albitized plagioclase and in veins. Heulandite occurs in the
upper kilometer of section and laumontite in the upper 2.5 km of section. Stewart
and Page (1974) reported early-formed calcareous concretions contain calcic plagioclase and no zeolites, whereas the surrounding sandstones have been altered to laumontite and albitized plagioclase.

British Columbia - Cretaceous to Eocene rocks. Read and Eisbacher (1974) described regional zeolitization of the continental Sustut Group in the Intermontaine
Belt (Figure 7). The Group consists of 2.3 km of interbedded volcanogenic and non-
volcanogenic sandstones. They reported that the mineral distribution is unrelated to
". . . volumetrically insignificant sills and plutons of granodiorite in the southeastern part of the basin" and that variations in silica activity controlled the
mineral distribution pattern.

Volcanic glass in tuffs has been replaced mainly by heulandite or less commonly
albite and/or analcime. Chiefly albite and quartz with minor heulandite fill pore
space of tuffaceous sandstones. The pore spaces of nontuffaceous sandstones are
filled by quartz-albite or laumontite. The laumontite appears to be restricted to
non-tuffaceous sandstones, and Read and Eisbacher (1974) postulated that calcium for
laumontite was supplied by dissolution of shell material.

Washington to Alaska - Tertiary rocks. Galloway (1974) described the widespread
occurrence of laumontite as a pore-filling cement in subsurface samples from Tertiary
basins ranging from Eocene to Pliocene age (Figure 7). Sandstones of the basins are
rich in plagioclase and volcanic rock fragments. Three cementation stages are recognized: (1) local, early calcite pore-filling cement, (2) authigenic clay rims around
detrital grains, and (3) authigenic clay and/or laumontite pore-filling cement (Figure
8). Galloway advocated temperature as a prime control on cement distribution but
recognized that there was widespread overlap. Assuming a 20°C surface temperature
and using thermal gradients given by Galloway for the area, the shallowest laumontite
occurrences would correspond to a temperature of about 55°-84°C.

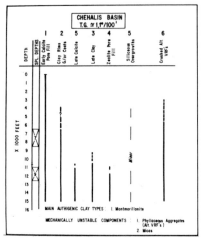

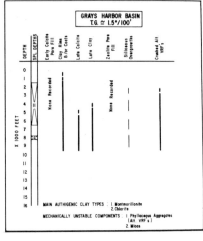

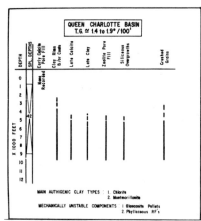

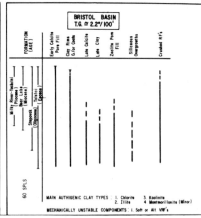

Figure 8. Distribution of cements, including laumontite, in four subsurface basins of western North America (from Galloway, 1974). See Figure 7 for locality of basins.

114

Washington - Tertiary rocks. Stewart (1974) described regional development of laumontite as a cement, in veins, and as a replacement of calcic plagioclase in marine Eocene sandstones of the Western Olympic Peninsula (Figure 7). Intermediate to basic rock fragments are abundant in the sandstones. Laumontite is more abundant in coarser grained sandstones rich in volcanic detritus than in finer grained, more feldspathic sandstones. Calcite-cemented sandstones are unzeolitized. Stewart estimated greater than 3 km of strata exposed, but burial depths are unknown. The adjacent area to the east contains prehnite-pumpellyite metamorphic rocks (Hawkins, 1967).

Oregon - Triassic and Jurassic rocks. Dickinson (1962a) and Brown and Thayer (1963) noted zeolitization in the Aldrich Mountain area of central Oregon (Figure 7). In both areas rocks are complexly folded in part, and small diorite porphyries intrude the area described by Brown and Thayer. Dickinson (1962a) studied 5.3 km of Jurassic marine sediments of andesitic origin and recognized: "(1) crystallization of volcanic glass to heulandite and associated minerals, including chlorite and celadonite; (2) local conversion of early-formed zeolites to laumontite; and (3) widespread decomposition of plagioclase to albite plus one or more hydrous Ca-bearing minerals including pumpellyite, prehnite, and laumontite." In addition, a rhyodacitic vitric tuff altered to heulandite was locally replaced by laumontite at some localities and by albite, quartz, and minor K-feldspar at others (Dickinson, 1962b). Similar relations have been observed in Triassic tuffs of New Zealand (Boles and Coombs, 1975). Dickinson did not recognize any consistent mineral distribution either with respect to stratigraphy or structure but emphasized the availability of water as a reaction control.

Brown and Thayer (1963) studied an area immediately east of Dickinson's area where an estimated 12-15 km of marine upper Triassic and lower Jurassic strata are exposed. This sequence is highly volcanogenic and contains tuffs, tuffaceous sandstones, and basaltic lavas. Albitization increases with stratigraphic age, but other mineral distributions were apparently controlled by the initial character of the rock. Thus, laumontite, prehnite, and pumpellyite are restricted to rocks partly or entirely of volcanogenic origin. Acidic vitric tuffs are commonly altered to albite-quartz or laumontite-quartz assemblages. Pumpellyite occurrences are restricted to basaltic lavas or basic rock fragments. Brown and Thayer (1963) also noted that zeolitization and plagioclase alteration was inhibited by the presence of calcite cement.

California - Cretaceous rocks. Dickinson et al. (1969) described alteration of the Great Valley sequence where some 10.7 km of section is exposed and an estimated

(Dickinson and others, 1969)

(Murata and Whiteley, 1973)

(Merino, 1975)

(Surdam and Hall, 1968)

(Castano and Sparks, 1974)

0   400 km

Figure 9. Index map of selected zeolite
occurrences in low-grade metamorphic rocks
of California. See text for description
of localities.

1.5 km of cover has been removed (Figure 9). The sequences form a steeply dipping, but
orderly, homoclinal succession. Sandstones commonly contain volcanic rock clasts (15-
45%), feldspars (20-40%), and quartzose clasts (25-55%). Increasing albitization of
plagioclase and chloritization of biotite with inferred burial depth has been recognized. Alteration was inhibited by the presence of carbonate cement. Laumontite is
common in the lower part of the section where it replaces plagioclase.

California - Tertiary rocks. Sporadic occurrences of zeolites in probable low-
grade metamorphic settings have been described at several localities (Figure 9). From
subsurface cores, Castaño and Sparks (1974) reported laumontite in Eocene-Miocene sandstones of the Tejon area at depths greater than 2.9 km, where corrected well temperatures are greater than or equal to 104°C and the fluid pressure is equal to 317 bars.
The zeolite occurs in arkosic sandstones as a cement and as a replacement of detrital
grains including mollusc shells. Plagioclase is more albitic (An_2-An_{15}) in the laumontite zone than stratigraphically above (An_2-An_{40}).

Laumontite occurs as a cement and more rarely as a replacement of plagioclase in
subsurface cores of Miocene volcanogenic sandstones from Kettleman North Dome (Merino,
1975). The laumontite is restricted to sandstones containing volcanic rock fragments.
From reconstructed burial thicknesses and present geothermal gradients, Merino (1975)
estimated about 105°C for the laumontite-bearing interval.

116

In surface outcrops, mordenite, stilbite, heulandite, and laumontite have been identified in Miocene sandstones and tuffs of the Coast Ranges. Mordenite is associated with clinoptilolite as a replacement of volcanic glass in tuffs of the marine Obispo Formation (Surdam and Hall, 1968) and in marine volcanogenic sandstones of the Briones Sandstone (Murata and Whiteley, 1973). Veins of laumontite and stilbite, extensive laumontite cements, and partial replacement of plagioclase by laumontite also occur in units of the Briones Sandstone. Plagioclase associated with laumontite is more albitic (oligoclase-albite) than in laumontite-free beds (andesine-oligoclase). The estimated depth of burial of the Briones Sandstone at the laumontite localities is less than 1 kilometer (Madsen and Murata, 1970). These workers also report a similar occurrence of laumontite in the marine Vaqueros Sandstone, containing very few volcanic rock fragments.

LITHOLOGY, CHEMISTRY, AND STABILITY OF ZEOLITE SPECIES

Heulandite-group Minerals

Lithology. Heulandite and clinoptilolite occur as a direct or indirect replacement of volcanic glass, either as pseudomorphs after glass shards or as an interstitial cement (Figure 10). They rarely replace other detrital phases such as feldspar. Heulandite-group minerals commonly coexist with phyllosilicates such as chlorite, smectite, celadonite, illite, and mixed-layer varieties. The clay usually forms a thin rim around the zeolite crystals.

The zeolite occurs as lath-shaped crystals, commonly between 10 and 100 μm in length, and exhibits parallel extinction along cleavage traces. The crystals may be length slow or length fast, but when samples are prepared in a hot mounting medium, crystals with Si/Al ratios equal to or greater than 3.57 are length slow, whereas crystals with lower Si/Al ratios are length fast (see Boles, 1972).

Heulandite-group minerals vary in abundance from a few percent as a cavity filling in sandstones to as much as 60% in altered vitric tuffs. In the latter rocks, minute, reddish iron oxide inclusions are common in the zeolite crystals, presumably by-products of the alteration of volcanic glass.

Chemistry. There are few quantitative data on the chemistry of heulandite-group minerals in low-grade metamorphic terraines. In the Southland Syncline of New Zealand, heulandites (Si/Al ratio <4.0) predominate, but clinoptilolites (Si/Al $\pm$ > 4.0) are also present (Boles, 1972; Boles and Coombs, 1975). The zeolites are generally sodium-poor

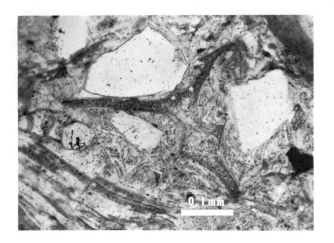

Figure 10a. Photomicrograph of heulandite pseudomorph after tricuspate-shaped glass shard. Dark inclusions in pseudomorph are iron oxides. Plain light. Triassic vitric-crystal tuff, Hokonui Hills, New Zealand.

Figure 10b. Photomicrograph of coarse-grained heulandite cement in volcanogenic sandstone. Plain light. Heulandite crystals have well-developed cleavage and crystals are rimmed by a pale-green phyllosilicate. Triassic volcanogenic sandstone, Hokonui Hills, New Zealand.

and calcium-rich (Figure 11). Magnesium ions average about 15% of total Ca + Mg ions. No systematic variation in zeolite composition with depth of burial has been found. Pyroclastic feldspars coexisting with clinoptilolites are more sodic than those associated with heulandites, implying that the glass precursors of the former species was more acidic than those of the latter (Boles and Coombs, 1975).

In Japan, sodic clinoptilolites predominate in such rocks, but calcic varieties have also been recognized (Utada, 1970; Minato and Utada, 1970; see Figure 11). Heulandite is much less common. Several Japanese studies have shown that heulandite or calcic clinoptilolite persists to inferred higher temperatures than alkalic clinoptilolite (Seki et al., 1969b; Utada, 1971; Iijima and Utada, 1972).

Stability. There have been no hydrothermal studies of heulandite-group minerals which demonstrated equilibrium with respect to another phase, and it is conceivable that members of this mineral group form as metastable phases (cf. Coombs, 1971). From the above case examples, it is clear that considerable overlap exists between the stability regions of heulandite and other minerals, but it is also clear that given sufficient burial depths (or temperatures) and appropriate pore-fluid chemistry,

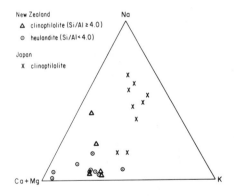

Figure 11. Na-(Ca+Mg)-K plot of compositions of heulandite-group minerals from low-grade metamorphic rocks. New Zealand data from Boles (1972) and Boles and Coombs (1975). Japan data from Utada (1970) and Minato and Utada (1971).

heulandite-group minerals are transformed to other phases. Reactions involving heulandite will also be discussed below, as the zeolite is an important precursor for a
number of phases.

Analcime

Lithology. Analcime occurs in a similar mode as heulandite-group minerals, i.e.,
as a direct or indirect product from volcanic glass of widely ranging composition
(Figure 12a). In a number of analcime tuffs from New Zealand, the zeolite forms pseudomorphs after heulandite-group minerals (Figure 12b). It rarely replaces plagioclase or
other detrital phases.

Analcime in cavities occurs as iscositetrahedral crystals ranging from 5 to 50 μm
across. The crystals are usually isotropic to nearly isotropic. Like heulandite, the
rocks richest in analcime are altered vitric tuffs. Analcime can make up as much as
60% of such rocks.

Chemistry. Analcime in low-grade metamorphic terraines is generally sodium rich,
although specimens with appreciable potassium substitution (up to 10 mole percent of
total Na+K+Ca+Mg) have been reported (Utada, 1970). Surdam (1973) and Seki (1973) reported extensive solid solution between analcime and wairakite. Analcime from low-
grade metamorphic rocks of New Zealand have silica/alumina ratios of about 2.4, whereas
in Japan they are slightly more siliceous (average 2.7). No relationship has been
found between depth of burial and analcime composition in either area (Coombs and
Whetten, 1969; Utada, 1970; Boles, 1971).

Stability. The hydrothermal stability of the analcime-albite system has been
investigated by Coombs et al. (1959), Campbell and Fyfe (1965), and Liou (1971a). The
upper thermal stability limit of analcime with respect to albite is about 200°C
(Figure 13). A lower temperature limit is expected when equilibrium with respect to
low albite is considered (see Liou, 1971a).

Figure 12a. Photomicrograph of vein of albite and trapazohedron-shaped analcime crystals (isotropic) in a Triassic vitric-crystal tuff altered to analcime
and quartz, Hokonui Hills, New Zealand. Crossed nicols.

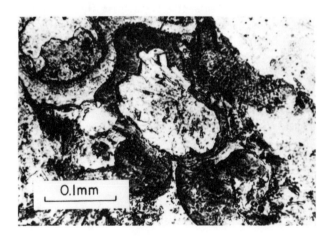

igure 12b. Lath-shaped analcime pseudomorphs after heulandite-group minerals in a
Triassic vitric-crystal tuff altered to analcime and quartz, Hokonui
Hills, New Zealand. Plain light. Pseudomorphs are embayed in chlorite.

121

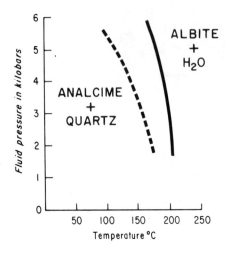

Figure 13. P_{fluid}-T diagram for the experimentally-determined reaction: analcime + quartz = intermediate albite + H_2O (solid line) and the calculated curve (dashed line) for the reaction: analcime + quartz = low albite + H_2O. Reaction curves for $P_{fluid} = P_{total}$. (After Liou, 1971a).

Laumontite

Lithology. Laumontite occurs as an indirect replacement of volcanic glass commonly through a zeolite precursor such as heulandite or analcime (Figure 14a). It commonly occurs as a cement and in some places as grains with no obvious volcanic precursor (see Vine, 1969). Laumontite has also been reported to replace fossil shell (Coombs, 1954; Castaño and Sparks, 1974).

Laumontite is commonly coarser grained (crystals up to 2 mm across in some rocks) and more abundant than other zeolites in low-grade metamorphic rocks. Laumontite commonly makes up from 30-80% of altered vitric tuffs. It is usually length slow along cleavage traces, although Coombs (1952) noted that the dehydrated leonhardite variety can be length fast.

Unlike most other zeolites, laumontite replaces plagioclase. The plagioclase has usually been albitized (Figure 14b), and presumably, albitization and laumontitization of the plagioclase are concurrent reactions.

In rocks where andesine plagioclase is albitized and laumontite forms as a cement, the reaction has the form:

$$NaAlSi_3O_8 \cdot CaAl_2Si_2O_8 + 3SiO_2 + 2H_2O + Na^+ =$$

andesine $\qquad$ quartz

$$2NaAlSi_3O_8 + 0.5CaAl_2Si_4O_{12} \cdot 4H_2O + 0.5Ca^{+2}$$

albite $\qquad$ laumontite

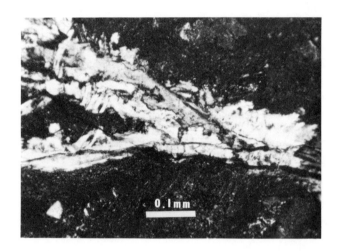

Figure 14a. Photomicrograph of laumontite (light color) as a vein and partial replacement of a Triassic heulandite-altered vitric tuff. Crossed nicols. Heulandite and dark green phyllosilicates are in dark areas. Faint shard
outlines are visible in laumontite.

Figure 14b. Photomicrograph of laumontite (light color) replacing plagioclase in a
Triassic volcanogenic sandstone. Crossed nicols. Plagioclase is albitized and is outlined by dotted line.

Such a reaction can account for ratios of laumontite cement to albitized plagioclase approaching 1:2 as observed in New Zealand rocks (Boles and Coombs, 1977).

Laumontite-bearing sandstones commonly have a mottled appearance due to unequal distribution of laumontite and phyllosilicate cement (Figure 15). Such sandstones have been described from many areas in the world (Hoare et al., 1964; Martini and Vaugnat, 1968; Madsen and Murata, 1970; Boles, 1974; Stewart and Page, 1974).

Chemistry. There are few analyses of laumontite in low-grade metamorphic terraines but available data indicate that the zeolite is essentially free of sodium and potassium and is near the ideal composition, $CaAl_2Si_4O_{12} \cdot nH_2O$ (Boles, 1971; Surdam, 1973).

Stability. Liou (1971b) determined the equilibrium boundary between laumontite-stilbite, laumontite-lawsonite, and laumontite-wairakite (Figure 16). The upper pressure limit of laumontite with respect to lawsonite is about 3.4 kb, and the upper temperature limit with respect to wairakite is about 300°C.

The presence of laumontite is often taken to indicate low-grade metamorphic conditions, and thus its lower thermal stability limit is of interest. Seki et al. (1969b) reported the apparent formation of laumontite at 75 $\pm$ 5°C in the Katayama geothermal field, Japan. Castaño and Sparks (1974) described the first appearance of laumontite at 104°C in California. Numerous studies of surface outcrops have reported laumontite at apparent shallow burial depths (e.g., Gill, 1975; Otálora, 1964; Hoare et al., 1964; Vine, 1969; Surdam, 1973); however, in such cases the amount of removed overburden is not well known and/or geothermal gradients were high. Recent studies in Southland Syncline rocks of New Zealand indicated that laumontite may have formed at temperatures as low as 50°C (Boles and Coombs, 1977). Sands and Drever (1977) reported laumontite in deep-sea sediments; if this laumontite is authigenic, it may have formed at temperatures considerably lower than described above.

Less Common Zeolites

Wairakite. As is expected from experimental data (Figure 16), wairakite is restricted to areas of relatively high geothermal gradients, including geothermal areas and areas with intrusive igneous bodies (e.g., Steiner, 1955; Coombs et al., 1959; Seki, 1973). Wairakite is commonly associated with laumontite and in a number of cases is believed to be a product of the breakdown of laumontite. Little is known about the P-T stability fields of sodium-bearing wairakites as described by Surdam (1973) and Seki (1973).

Figure 15.   Laumontite-bearing mottled sandstones, Hokonui Hills, New Zealand. Mottled
texture due to unequal distribution of laumontite (light patches) and phyllosilicates (dark patches).

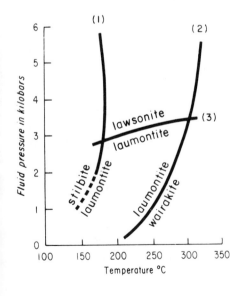

Figure 16. P_{fluid}-T diagram for the experimentally-determined reactions:

(1) stilbite = laumontite + 3 quartz + $3H_2O$;

(2) laumontite = wairakite + $2H_2O$;

(3) laumontite = lawsonite + 2 quartz + $2H_2O$;

at P_{fluid} = P_{total}. (After Liou, 1971 b,c).

Mordenite. Mordenite occurrences are restricted to Tertiary and younger sediments as replacements of volcanic glass. Mordenite is relatively common in low-grade metamorphic rocks of Japan (see Utada, 1970). It has also been reported at several localities in California in possible low-grade metamorphic settings (Surdam and Hall, 1968; Murata and Whiteley, 1973) and in the Soviet Union (Gogishvili, 1976). Mordenite commonly occurs in active geothermal areas as one of the earliest formed zeolites (Steiner, 1955; Seki et al., 1969b; Honda and Muffler, 1970; Kristmannsdottir and Tomasson, 1977). Mordenite is commonly associated with heulandite-group minerals, but has also been found with laumontite (Seki et al., 1971).

Experimental work of Hawkins et al. (1977) indicates that at a fixed fluid composition, mordenite forms at higher temperatures than clinoptilolite. Judging from these experimental data and many of the above occurrences, mordenite may be indicative of relatively high geothermal gradients in rocks of appropriate bulk composition and may have a higher thermal stability than heulandite-group minerals.

Stilbite. Stilbite is commonly found as a vein mineral or locally as a replacement product in volcanogenic rocks in association with heulandite-group minerals or laumontite. It has been described from New Zealand (Coombs, 1954), Japan (Seki et al., 1969a), California (Murata and Whiteley, 1973), and Iceland (Kristmannsdottir and Tomasson, 1977). Staples (1965) described the replacement and filling of fossil shells by stilbite and other zeolites as a result of hydrothermal solutions. Figure 16 shows the stability field of stilbite with respect to laumontite. Little is known about the composition of stilbite in low-grade metamorphic rocks, but a few analyses indicate it is dominantly calcium-rich and has a Si/Al ratio ranging from about 2.5 to 3.0 (Liou, 1971c; Boles, 1971).

Other zeolites. Erionite, chabazite, ferrierite and yugawaralite have been reported in Japan (see Utada, 1970). Erionite, chabazite, and phillipsite of possible hydrothermal origin occur in altered Tertiary tuffs of New Zealand (Sameshima, 1978). These zeolites are much less widespread than others mentioned above and appear to be restricted to Tertiary and younger rocks.

ZEOLITE REACTIONS

Volcanic glass and calcic plagioclase are the most important precursors for zeolites in low-grade metamorphic rocks. Early reactions involve hydration of volcanic glass to phyllosilicates including smectite, chlorite, celadonite, and mixed-layer clay.

Forming contemporaneously, and, in part, later than the clay, are zeolites such as heulandite and clinoptilolite. These zeolites may have had a zeolite precursor such as phillipsite (see zeolites in deep-sea sediment chapter), but unambiguous data are lacking.

Still later reactions involve the replacement of calcic plagioclase by laumontite, the penecontemporaneous albitization of the plagioclase, and the replacement of early-formed zeolites by less hydrous and less siliceous phases. Dehydration-desilicification reactions that have been documented include the following (see Dickinson, 1962b; Utada, 1971; Surdam, 1973; Boles and Coombs, 1975, 1977) (H = heulandite, L = laumontite, Pr = prehnite, An = analcime, Ab = albite, Pu = pumpellyite):

(1) $\overset{\text{H}}{Ca_3K_2Al_8Si_{28}O_{72}\cdot 23H_2O} \rightarrow \overset{\text{L}}{3CaAl_2Si_4O_{12}\cdot 4H_2O} + 10SiO_2 + 2KAlSi_3O_8 + 11H_2O$

(2) $3Ca^{++} + \overset{\text{H}}{Ca_3K_2Al_8Si_{28}O_{72}\cdot 23H_2O} \rightarrow \overset{\text{Pr}}{3Ca_2Al_2Si_3O_{10}(OH)_2} + 2KAlSi_3O_8 + 13SiO_2 + 17H_2O + 6H^+$

(3) $8Na^+ + \overset{\text{H}}{Ca_3K_2Al_8Si_{28}O_{72}\cdot 23H_2O} \rightarrow \overset{\text{An}}{8NaAlSi_2O_6\cdot H_2O} + 12SiO_2 + 2K^+ + 3Ca^{++}$

(4) $8Na^+ + \overset{\text{H}}{Ca_3K_2Al_8Si_{28}O_{72}\cdot 23H_2O} \rightarrow \overset{\text{Ab}}{8NaAlSi_3O_8} + SiO_2 + 23H_2O + 2K^+ + 3Ca^{++}$

(5) $\overset{\text{An}}{NaAlSi_2O_6\cdot H_2O} + SiO_2 \rightarrow \overset{\text{Ab}}{NaAlSi_3O_8} + H_2O$

(6) $\overset{\text{L}}{CaAl_2Si_4O_{12}\cdot 4H_2O} + Ca^{++} \rightarrow \overset{\text{Pr}}{Ca_2Al_2Si_3O_{10}(OH)_2} + SiO_2 + 2H_2O + 2H^+$

(7) $\overset{\text{L}}{2CaAl_2Si_4O_{12}4H_2O} + 2Ca^{++} \rightarrow \overset{\text{Pu}}{Ca_4(Fe,Mg)^{2+}Fe^{3+}Al_4Si_6O_{21}(OH)_7} + 2SiO_2 + 9H^+$
$+ (Fe,Mg)^{++} + Fe^{+++}$

In addition to the breakdown of early-formed zeolites, phyllosilicates are also modified or are consumed by reactions. For example, in reaction (1) some of the early-formed phyllosilicate associated with heulandite is converted to laumontite in vitric tuffs of New Zealand (Boles and Coombs, 1975).

A notable feature of reactions (3) and (4) is that they imply considerable mobility of Na^+, Ka^+, and Ca^{+2} ions. In fact, late-stage sodium metasomatism of calcic zeolites on widely varying scales is characteristic of low-grade metamorphic terraines in several areas (Coombs, 1954; Dickinson, 1962b; Utada, 1970; Boles and Coombs, 1975). Such re-placement would obviously have important consequences to the nature of higher grade metamorphic assemblages.

Unlike the rocks in more conventional metamorphic settings, zeolite-bearing, low-grade metamorphic rocks consists of heterogeneous mineral assemblages of relict detrital grains and early- and late-formed diagenetic-metamorphic minerals. The overlapping distribution patterns of zeolites in such terraines is better understood when one considers the possible complex interplay of factors controlling zeolite stability, discussed below.

Effect of Pressure

Load pressure appears to be one of the least significant variables affecting zeolite stability, as evidenced by the wide range of burial depths at which zeolites occur in low-grade metamorphic terraines. As pointed out by Zen (1974), most low-grade metamorphic reactions involve relatively small changes in volumes of solids and hence the effect of pressure should be minor.

Zeolite reactions in low-grade metamorphic rocks commonly involve dehydration. As pointed out by Coombs et al (1959) and by Zen (1974), fluid pressure/load-pressure ratios can markedly affect equilibrium boundaries of such reactions. For example, Zen (1974) calculated that the equilibrium for the reaction laumontite = lawsonite + 2 quartz + $2H_2O$ (see Figure 16) will be lowered at constant temperature of 200°C from 2.9 kb where $P_f = P_t$ to about $P_t = 2$ kb where $P_f = 1.5$ kb. The effect of varying P_f/P_t ratios may be marked on reactions where large quantities of water are evolved (e.g., heulandite → feldspar).

Conditions at which P_f is less than P_t are undoubtedly common in shallow burial settings. Essentially hydrostatic conditions (i.e., $P_f = 0.5P_t$) are found as deep as 5 km in some areas of the Gulf Coast Geosyncline (see Dickinson, 1953; Jones, 1969). Lower P_f/P_t ratios can develop in fractures than in the adjacent country rock, causing displacement of the equilibrium boundary between hydrous and less hydrous phases. In New Zealand, this effect can be demonstrated at several localities where zeolites in the country rock are replaced by less hydrous phases in fractures (Figures 12a, 14a; Boles and Coombs 1977). Where P_f/P_t ratios increase rapidly during burial of sediments, analogous to conditions in the Gulf Coast Geosyncline, considerable overlap in mineral distributions may result.

Effect of Temperature

Temperature is a very important factor in controlling zeolite stability, largely because the dehydration resulting from zeolite breakdown involves relatively large entropy changes. Zeolites such as members of the heulandite group and analcime clearly can form at essentially STP conditions, whereas zeolites such as laumontite and wairakite require elevated temperatures. The grossly overlapping mineral assemblages in many low-grade metamorphic terraines, however, demonstrates that factors other than temperature are also important.

128

The temperatures at which zeolites occur in burial metamorphic settings is usually inferred from the stratigraphic distribution of zeolites combined with an estimate of the geothermal gradient. Many of the zeolite occurrences described in this chapter are associated with thick piles of sediments deposited in basins near convergent plate margins. Aside from present structural complexities, true burial thickness may be considerably less than measured stratigraphic thickness, owing to shifts in the locus of deposition. Until sediment dispersal patterns and heat distribution in such basins are better understood, the effects of temperature on zeolite assemblages in these settings can not be properly evaluated.

The rise in temperature associated with increasing burial depth may undergo perturbations from hydration reactions involving zeolites. Surdam and Boles (1977) calculated that the hydration of andesine to laumontite in a sandstone with density 2.3 g/cc, 40% andesine, and an initial temperature of 60°C at 1.5 km of burial would raise the temperature of the rock by 40°C if the heat of the reaction is conserved. Zeolitization of volcanic glass should also evolve heat.

The effects of hydrothermal waters associated with igneous intrusions may be underestimated in many zeolitic low-grade metamorphic terraines. Although conventional contact metamorphic effects are usually restricted to a few meters of an intrusive body, oxygen isotope studies suggest that intrusives are. capable of heating porous and permeable country rocks to temperatures greater than 100°C at distances of at least three stock diameters from the intrusions (Taylor, 1971). Thus, in areas where intercalated lavas and intrusive rocks are present, zeolitization may have been influenced by emplacement of the igneous rocks as well as by "burial" effects.

Effect of Parent Material and Bulk Composition

Volcanic glass and calcic plagioclase are the most common precursor for zeolites in low-grade metamorphic terraines. The absence of zeolites in similar terraines, for example, the Salton Sea geothermal area (Muffler and White, 1969), may be due to the absence of volcanogenic parent materials. The initial composition of detrital glasses has been shown to affect the mineral assemblages and mineral composition (Brown and Thayer, 1963; Horne, 1968; Stewart, 1974; Boles and Coombs, 1977). For example, calcium heulandite and/or laumontite prevail in altered andesitic detritus of New Zealand, whereas authigenic albite predominates in altered rhyo-dacitic detritus in the same region. Similarly, where rhyo-dacitic volcanic detritus predominates, siliceous, alkali-rich zeolites predominate in Japan. Silica/alumina ratios of heulandite-group minerals may also be related to initial glass composition (see Boles and Coombs, 1975).

Effect of Fluid Chemistry

Pore-fluid chemistry can be an important control on mineral reactions in diagenetic low-grade metamorphic terraines. Initially, the fluid/solid ratios are high and fluids are relatively mobile. Several zeolitic reactions at fixed temperature and pressure may involve only a_{H_2O} and/or a_{SiO_2} in pore fluids, for example, (1) heulandite = laumontite + quartz + H_2O, and (2) laumontite = wairakite + H_2O. Other documented reactions, for example those involving alteration of calcic zeolites to analcime or albite, are dependent not only on pressure and temperature, but on the activities of H_2O, silica, sodium, and calcium in pore fluids. As shown in Figure 17a, at constant temperature and pressure heulandite is converted to analcime or albite depending on the initial heulandite composition and on the nature and extent of change of activities of silica, water, and Na^+/Ca^{+2} in the pore fluid.

Zen (1961) and Thompson (1971) pointed out that P_{CO_2} may be an important control on zeolite equilibria where carbonates are involved in zeolitic reactions. Thus, clay-carbonate assemblages can replace zeolite assemblages in areas of high P_{CO_2}. The effect is well demonstrated in hydrothermal $-$ altered volcanic rocks of New Zealand. In the Ohaki-Broadlands area, P_{CO_2} values are much higher than at Wairakei; clay-carbonate assemblages are much more common at Wairakei (Browne and Ellis, 1970). It should also

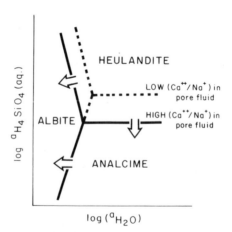

Figure 17a. Activity diagram at constant temperature, pressure, and activity of H_2O showing phase relations for albite, analcime, and heulandite ($Ca_{3.5}NaAl_8 Si_{28}\cdot24H_2O$). Note that equilibrium boundary will shift as a function of varying Ca/Na activity ratios in pore fluids. Arrows show direction of common reactions in low-grade metamorphic rocks. [Error: The slope of the heulandite-analcime phase boundary should be +2/3.]

130

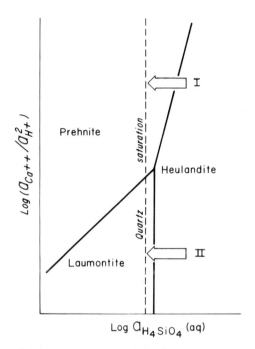

Figure 17b. Activity diagram at constant temperature, pressure, and activity of H_2O showing phase relations for laumontite, $CaAl_2Si_4O_{12} \cdot 4H_2O$; heulandite, $CaAl_2Si_7O_{18} \cdot 6H_2O$; and prehnite, $Ca_2Al_2Si_3O_{14}(OH)_2$. Arrows show direction of the most common reactions between the phases in rocks of the Hokonui Hills, New Zealand. (After Boles and Coombs, 1977.)

be noted that clay-carbonate-zeolite assemblages are very common in low-grade metamorphic terraines; thus, the presence of clay and carbonates does not necessarily preclude zeolite assemblages.

Effect of Permeability

Permeability is an obvious control on reactions where the transfer of ions to and from the reaction site is required. Numerous studies have shown that early-formed calcite cements inhibit the formation of zeolite cements and albitization of plagioclase (Horne, 1968; Dickinson et al., 1969; Stewart and Page, 1974). In addition, albitization of plagioclase, a reaction which is an important source of laumontite cement, is more complete in coarse-grained sediments where permeabilities are presumably highest (see Boles and Coombs, 1977).

Effect of Kinetics

Reaction kinetics between zeolites and between zeolites and less hydrous phases are poorly understood. Hay (1966) pointed out that with increasing age zeolitic assemblages become simpler and that in early Paleozoic rocks they have been replaced by feldspar. In a general way, this is true of low-grade metamorphic assemblages, i.e., laumontite, heulandite-group minerals, and analcime are characteristic of Mesozoic and late Paleozoic rocks, whereas much more diverse assemblages can be found in younger rocks.

Many of the reactions which have been demonstrated from field and petrographic evidence are not isochemical in that they involve the transfer of ions to and from the reaction site. Reaction kinetics are therefore functions of the diffusion rates of ions and/or the mobility of pore fluids. To date, there are few quantitative data on such processes in geologic settings.

REFERENCES

Boles, J.R. (1971) Stratigraphy, petrology, mineralogy, and metamorphism of mainly Triassic rocks, Hokonui Hills, Southland, New Zealand: unpublished Ph.D. thesis, University of Otago, Dunedin, New Zealand, 406 pp.

Boles, J.R. (1972) Composition, optical properties, cell dimensions, and thermal stability of some heulandite-group zeolites: Am. Mineral. 57, 1463-1493.

Boles, J.R. (1974) Structure, stratigraphy, and petrology of mainly Triassic rocks, Hokonui Hills, Southland, New Zealand: N. Z. J. Geol. Geophys. 17, 337-374.

Boles, J.R., and Coombs, D.S. (1975) Mineral reactions in zeolitic Triassic tuff, Hokonui Hills, New Zealand: Geol. Soc. Am. Bull. 86, 163-173.

Boles, J.R., and Coombs, D.S. (1977) Zeolite facies alteration of sandstones, Southland Syncline, New Zealand: Am. J. Sci. (in press).

Brown, C.E., and Thayer, T.P. (1963) Low-grade mineral facies in upper Triassic and lower Jurassic rocks of the Aldrich Mountains, Oregon: J. Sediment. Petrol. 33, 411-425.

Browne, P.R.L., and Ellis, A.J. (1970) The Ohaki-Broadlands area, New Zealand: Am. J. Sci. 269, 97-131.

Campbell, A.S., and Fyfe, W.S. (1965) Analcime-albite equilibria: Am. J. Sci. 263, 807-816.

Castaño, J.R., and Sparks, D.M. (1974) Interpretation of vitrinite reflectance measurements in sedimentary rocks and determination of burial history using vitrinite reflectance and authigenic minerals: Geol. Soc. Am. Spec. Paper 153, 31-52.

Coombs, D.S. (1952) Cell size, optical properties and chemical composition of laumontite and leonhardite: Am. Mineral. 37, 812-830.

Coombs, D.S. (1954) The nature and alteration of some Triassic sediments from Southland, New Zealand: Trans. Roy. Soc. New Zealand 82, 65-109.

Coombs, D.S. (1971) Present status of the zeolite facies: In, Advances in Chemistry Series No. 101, "Molecular Sieve Zeolites - I," Am. Chem. Soc., Washington, D. C., 317-327.

Coombs, D.S., Ellis, A.J., Fyfe, W.S., and Taylor, A.M. (1959) The zeolite facies, with comments on the interpretation of hydrothermal syntheses: Geochim. Cosmochim. Acta 17, 53-107.

Coombs, D.S., and Whetten, J.T. (1967) Composition of analcime from sedimentary and burial metamorphic rocks: Geol. Soc. Am. Bull. 78, 269-282.

Dickinson, G. (1953) Geologic aspects of abnormal reservoir pressures in Gulf Coast Louisiana: Am. Assoc. Petrol. Geol. Bull. 37, 410-432.

Dickinson, W.R. (1962a) Petrology and diagenesis of Jurassic andesitic strata in central Oregon: Am. J. Sci. 260, 481-500.

Dickinson, W.R. (1962b) Metasomatic quartz keratophyre in central Oregon: Am. J. Sci. 260, 249-266.

Dickinson, W.R., Ojakangas, R.W., and Stewart, R.J. (1969) Burial metamorphism of the late Mesozoic Great Valley sequence, Cache Creek, California: Geol. Soc. Am. Bull. 80, 519-526.

Eskola, P. (1920) The mineral facies of rocks: Norsk Geol. Tidsskr. 6, 143-194.

Fyfe, W.S., Turner, F.J., and Verhoogen, J. (1958) Metamorphic reactions and metamorphic facies: Geol. Soc. Am. Mem. 73, 259 pp.

Galloway, W.E. (1974) Depositional and diagenetic alteration of sandstone in northeast Pacific arc-related basins: Implications for graywacke genesis: Geol. Soc. Am. Bull. 85, 379-390.

Gill, E.D. (1957) Fossil wood replaced by laumontite near Cape Patterson, Victoria: Proc. Roy. Soc. Victoria 69, 33-35.

Gogishvili, V.G. (1976) Mordenite and clinoptilolite of the Transcaucasus: Zeolite '76 Conf., Tucson, AZ, June 1976, 27-28 (abstract).

Hawkins, D.B., Sheppard, R.A., and Gude, A.J., 3rd (1978) Hydrothermal synthesis of clinoptilolite and comments on the assemblage phillipsite-clinoptilolite-mordenite: In, L.B. Sand and F.A. Mumpton, Eds., Natural Zeolites: Occurrence, Properties, Use, Pergamon Press, Elmsford, NY, 337-343.

Hawkins, J.W., Jr., (1967) Prehnite-pumpellyite facies metamorphism of a graywacke-shale series, Mount Olympus, Washington: Am. J. Sci. 265, 798-818.

Hay, R.L. (1966) Zeolites and zeolitic reactions in sedimentary rocks: Geol. Soc. Am. Spec. Paper 85, 130 pp.

Hoare, J.M., Condon, W.H., and Patton, W.W., Jr. (1964) Occurrence and origin of laumontite in Cretaceous sedimentary rocks in western Alaska: U.S. Geol. Surv. Prof. Pap. 501-C, C74-C78.

Honda, S., and Muffler, L.J.P. (1970) Hydrothermal alteration in core from research drill hole Y-1, Upper Geyser Basin, Yellowstone National Park, Wyoming: Am. Mineral. 55, 1714-1737.

Horne, R.R. (1968) Authigenic prehnite, laumontite and chlorite in the Lower Cretaceous sediments of south-eastern Alexander Island: Br. Antarct. Surv. Bull. 18, 1-10.

Iijima, A., and Utada, M. (1972) A critical review on the occurrence of zeolites in sedimentary rocks in Japan: Japan J. Geol. Geogr. 42, 61-83.

Jolly, W.T. (1970) Zeolite and prehnite-pumpellyite facies in south central Puerto Rico: Contrib. Mineral. Petrol. 27, 204-224.

Jones, P.H. (1969) Hydrodynamics of geopressure in the northern Gulf basin: J. Petrol. Tech., July issue, 803-810.

Kossovskaya, A.G., and Shutov, V.D. (1961) The correlation of zones of regional epi- genesis and metagenesis in terrigneous and volcanic rocks: Akad. Nauk. SSSR Dok- lady, Earth Sci. Sec. 139, 677-680.

Kristmannsdottir, H., and Tomasson, J. (1978) Zeolite zones in geothermal areas in Iceland: In, Sand, L.B. and Mumpton, F.A., Eds., Natural Zeolites: Occurrence, Properties, Use, Pergamon Press, Elmsford, N. Y., 277-284.

Liou, J.G. (1971a) Analcime equilibria: Lithos 4, 389-402.

Liou, J.G. (1971b) P-T stabilities of laumontite-wairakite, lawsonite, and related minerals in the system $CaAl_2Si_2O_8-SiO_2-H_2O$: J. Petrol. 12, 379-411.

Liou, J.G. (1971c) Stilbite-laumontite equilibrium: Contrib. Mineral. Petrol. 31, 171-177.

Madsen, B.M., and Murata, K.J. (1970) Occurrence of laumontite in Tertiary sandstones of the central Coast Ranges, California: U.S. Geol. Surv. Prof. Paper 700-D, D188-D195.

Martini, J., and Vaugnat, M. (1968) Étude pétrographique des Grès de Taveyanne entre Arve et Giffre (Haute-Savoie, France): Schweiz. Mineral. Petrogr. Mitt. 48, 539-654.

Merino, E. (1975) Diagenesis in Tertiary sandstones from Kettleman North Dome, Cali- fornia. I. Diagenetic mineralogy: J. Sediment. Petrol. 45, 320-336.

Minato, H., and Utada, M. (1969) Zeolite: In, Clays of Japan, Geol. Surv. of Japan Publ., 121-134.

Minato, H., and Utada, M. (1971) Clinoptilolite from Japan: Advances in Chemistry Series No. 101, Molecular Sieve Zeolites - I, Am. Chem. Soc., Washington, D. C., 311-316.

Mossman, D.J., and Bachinski, D.J. (1972) Zeolite facies metamorphism in the Silurian- Devonian fold belt of northeastern New Brunswick: Can. J. Earth Sci. 9, 1703-1709.

Muffler, L.J.P., and White, D.E. (1969) Active metamorphism of upper Cenozoic sedi- ments in the Salton Sea geothermal field and the Salton Trough, southeastern California: Geol. Soc. Am. Bull. 80, 157-181.

Murata, K.J., and Whiteley, K.R. (1973) Zeolites in the Miocene Briones Sandstone and related formations of the central Coast Ranges, California: J. Res. U.S. Geol. Surv. 1, 255-265.

Otálora, G. (1964) Zeolites and related minerals in Cretaceous rocks of east-central Puerto Rico: Am. J. Sci. 262, 726-734.

Read, P.B., and Eisbacher, G.H. (1974) Regional zeolite alteration of the Sustut Group, north-central British Columbia: Can. Mineral. 12, 527-541.

Sameshima, T. (1978) Zeolites in tuff beds of the Waitemata Group, Auckland Province, New Zealand: In, Sand, L.B. and Mumpton, F.A., Eds., Natural Zeolites: Occur- rence, Properties, Use, Pergamon Press, Elmsford, NY, 309-317.

Sands, C.D., and Drever, J.I. (1978) Authigenic laumontite in deep-sea sediments: In, Natural Zeolites: Occurrence, Properties, Use, Sand, L.B. and Mumpton, F.A., Eds., Pergamon Press, Elmsford, NY 269-275.

Seki, Y. (1973) Distribution and modes of occurrence of wairakites in the Japanese Island Arc: J. Geol. Soc. Japan 79, 521-527.

Seki, Y., Oki, Y., Matsuda, T., Mikami, K., and Okumura, K. (1969a) Metamorphism in the Tansawa Mountains, central Japan: J. Japan. Assoc. Mineral. Petrol. Econ. Geol. 61, 1-75.

Seki, Y., Onuki, H., Okumura, K., and Takashima, I. (1969b) Zeolite distribution in the Katayama geothermal area of Japan: Japan. J. Geol. Geogr. 40, 63-79.

Staples, L.W. (1965) Zeolite filling and replacement in fossils: Am. Mineral. 50, 1796-1801.

Steiner, A. (1955) Hydrothermal rock alteration: Dep. Sci. Indus. Res. Bull. N.Z. 117, 21-26.

Stewart, R.J. (1974) Zeolite facies metamorphism of sandstone in the western Olympic Peninsula, Washington: Geol. Soc. Am. Bull. 85, 1139-1142.

Stewart, R.J., and Page, R.J. (1974) Zeolite facies metamorphism of the late Cretaceous Nanaimo Group, Vancouver Island and Gulf Islands, British Columbia: Can. J. Earth Sci. 11, 280-284.

Surdam, R.C. (1973) Low-grade metamorphism of tuffaceous rocks in the Karmutsen Group, Vancouver Island, British Columbia: Geol. Soc. Am. Bull. 84, 1911-1922.

Surdam, R.C., and Hall, C.A. (1968) Zeolitization of the Obispo Formation, Coast Ranges of California (abstr.): Geol. Soc. Am. Spec. Paper 101, 338 pp.

Surdam, R.C., and Boles, J.R. (1977) Diagenesis of volcanogenic sandstones (abstr.): Program, Am. Assoc. Petrol. Geol. Ann. Meet., Denver, Colorado.

Taylor, H.P., Jr. (1971) Oxygen isotope evidence for large-scale interaction between intrusions, Western Cascade Range, Oregon: J. Geophys. Res. 76, 7855-7874.

Thompson, A.B. (1971) P_{CO_2} in low-grade metamorphism; zeolite carbonate, clay mineral, prehnite relations in the system $CaO-Al_2O_3-SiO_2-CO_2-H_2O$: Contrib. Mineral. Petrol. 33, 145-161.

Utada, M. (1970) Occurrence and distribution of authigenic zeolites in the Neogene pyroclastic rocks in Japan: Sci. Papers College of General Education, Univ. of Tokyo 20, 191-262.

Utada, M. (1971) Zeolitic zoning of the Neogene pyroclastic rocks in Japan: Sci. Papers College of General Education, Univ. of Tokyo 21, 189-221.

Vine, J.D. (1969) Authigenic laumontite in arkosic rocks of Eocene age in the Spanish Peaks area, Las Animas County, Colorado: U.S. Geol. Surv. Prof. Paper 650-D, D80-D83.

Zaporozhtseva, A.S. (1960) On the regional development of laumontite in Cretaceous deposits of Lena Coal Basin: Izv. Akad. Nauk SSSR, Ser. Geol., no. 9, English translation, 52-59.

Zaporozhtseva, A.S., Vishnevskaya, T.N., and Dubar, G.P. (1961) Successive change in calcium zeolites through vertical sections of sedimentary strata: Dok. Akad. Nauk SSSR, Earth Sci. Sec. 141, 448-451.

Zen, E-An (1961) The zeolite facies: an interpretation: Am. J. Sci. 259, 401-409.

Zen, E-An (1974) Burial metamorphism: Can. Mineral. 12, 445-455.

Chapter 7

ZEOLITES IN DEEP-SEA SEDIMENTS [*]

James R. Boles

INTRODUCTION

Zeolites are important diagenetic minerals in deep-sea sediments. They may have a significant influence on pore-water chemistry and on the mass balance of elements in this environment (Glaccum and Boström, 1976; Kastner and Stonecipher, 1978). Because zeolites are actively forming in deep-sea sediments, and burial histories, thermal gradients, sediment ages, and pore-water chemistries are reasonably well known, much can be learned about the zeolitization process by studying their occurrence in this environment.

Phillipsite and clinoptilolite are the two most abundant zeolites in deep-sea sediments and make up as much as 80% by weight of the sediment (Czyscinski, 1973). Phillipsite was first recognized in the late 1800s (Murray and Renard, 1891), but clinoptilolite was not recognized until the mid 1960s (Biscaye, 1965; Hathaway and Sachs, 1965). Since their discovery, phillipsite and clinoptilolite have been found to be widespread and frequently abundant diagenetic constituents of deep-sea sediments. Most of our recent knowledge of zeolites in this environment comes from drill cores of the Deep Sea Drilling Project. Phillipsite has been recognized by x-ray diffraction analysis in more than 320 samples of DSDP cores and clinoptilolite in more than 650 samples. Phillipsite is also present in about 10% of the clinoptilolite-bearing samples. Other zeolites recognized in deep-sea sediments in approximate order of decreasing abundance include analcime, erionite, and laumontite. The occurrences have been summarized by Kolla and Biscaye (1973), Cronan (1974), Stonecipher (1976), Kastner and Stonecipher (1978), and Boles and Wise (1978), and the reader is referred to these papers.

— — — — — — — — — — — — —

[*]
   Editor's note: See also Chapter 4, "Zeolites" in Volume 6 of *Reviews in Mineralogy* (formerly *"Short Course Notes")* published by the Mineralogical Society of America Washington, D.C., 1979.

OCCURRENCE AND MINERALOGY OF PHILLIPSITE AND CLINOPTILOLITE

Age of Sediment

Probably the most striking fact about phillipsite and clinoptilolite occurrences[1] is their relationship to sediment age (Stonecipher, 1976; Boles and Wise, 1978). Phillipsite is most abundant in Miocene and younger sediments whereas clinoptilolite is most abundant in Eocene and older sediments (Figure 1a). The percentage of phillipsite occurrences increases from the Recent to about the Miocene and then rapidly decreases; whereas, the percentage of clinoptilolite occurrences show a progressive increase at least to the Cretaceous. This "transition" downward from phillipsite to clinoptilolite has been observed at a number of individual drill sites (Stonecipher, 1976). Not only does the number of clinoptilolite occurrences increase with geologic age, but also its abundance in samples increases with geologic age (Figure 1b).

Several studies have shown that phillipsite nucleates at or within centimeters of the sediment-water interface and continues to grow in the sediment column. Growth periods have been estimated to range from 150,000 years (Czyscinski, 1973) to at least 10^6 years (Bernat et al., 1970). The increase in phillipsite occurrences from Recent to Miocene age (Figure 1a) suggests growth periods of greater than 10^7 years. In older sediments, phillipsite crystals are frequently etched suggesting dissolution (Rex, 1967; Kastner and Stonecipher, 1978). This may explain the paucity of phillipsite in Cretaceous sediments.

Type of Sediments

About 70-80% of phillipsite and clinoptilolite occurrences are in fine-grained pelagic sediments including brown clays, nannofossil oozes, calcareous oozes, or siliceous oozes (Figure 2). These sediment types were formed at relatively low sedimentation rates ranging from 0.2-2 m/10^8 years for clays to 10-30 m/10^6 years for calcareous oozes (Heezen and Hollister, 1971). As shown in Figure 2, phillipsite and clinoptilolite also occur in numerous other sediment types.

About 33% of the clinoptilolite occurrences recognized from bulk-rock x-ray diffraction analyses are in calcareous oozes, marls, and limestones, whereas 28% of the phillipsite occurrences are in this sediment type (Figure 2). Whether or not this

----

[1]"Occurrence" is used in this review to denote the presence of a mineral identified by x-ray diffraction in a bulk rock core sample and described in the Initial Reports of the Deep Sea Drilling Project vol. 1-35.

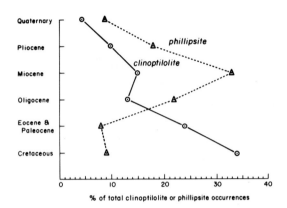

% of total clinoptilolite or phillipsite occurrences

Figure 1a. Distribution of deep-sea phillipsite and clinoptilolite samples with respect to age of sediment (modified from Boles and Wise, 1978). Eocene and Paleocene data have been combined owing to the paucity of data in Paleocene sediments. Data from bulk-rock x-ray diffraction analyses of DSDP cores (see Initial Reports of Deep Sea Drilling Project vol. 1-35). Data include 321 phillipsite occurrences and 649 clinoptilolite occurrences.

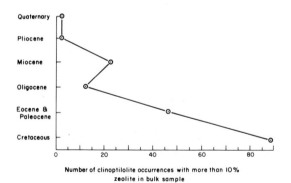

Number of clinoptilolite occurrences with more than 10% zeolite in bulk sample

Figure 1b. Distribution of deep-sea clinoptilolite samples containing greater than 10% clinoptilolite in the bulk rock (after Boles and Wise, 1978). See Figure 1a for data source.

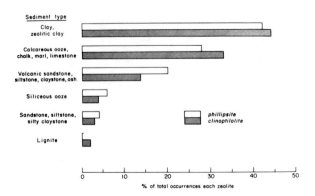

Figure 2. Distribution of deep-sea phillipsite with respect to sediment type (modified from Boles and Wise, 1978). See Figure 1a for data source.

difference is significant is uncertain, but Stonecipher (1976, 1978) noted that clinoptilolite is more frequently found in calcareous sediments than is phillipsite, if x-ray diffraction analyses of the 2-20 μm size fraction are included in the data. Stonecipher contends that the presence of carbonate may act as a catalyst for clinoptilolite crystallization.

Unusual occurrences include phillipsite within manganese nodules (Murray and Renard, 1891; Bonatti, 1963; Bonatti and Nayadu, 1965) and as a replacement of plagioclase (Bass et al., 1973). Clinoptilolite has also been described as large crystals in a lignite from the East Indian Ocean (Boles and Wise, 1978) and as a replacement of radiolarian tests (Berger and von Rad, 1972; Benson, 1972; Robinson et al., 1974).

One of the more surprising aspects of Figure 2 is that phillipsite and clinoptilolite are not commonly associated with volcanic-rich detritus although the association is more evident for phillipsite than clinoptilolite. There is an apparent inverse correlation between zeolite occurrences and volcanic ash beds in a number of cores (see Boles and Wise, 1978). In other cores (e.g., from DSDP Leg 34), phillipsite and clinoptilolite occur ". . . in almost pure nannofossil oozes in which it is difficult to find direct evidence of volcanic debris except in the existence of zeolites" (Bass, 1976, p. 619; also, see Kolla and Biscaye, 1973; Stonecipher, 1978). Minor amounts of volcanic glass may have once been present in many such pelagic sediments, but it subsequently dissolved by reaction with pore fluids. Nevertheless, there appears to be reasonably good evidence that volcanic glass is not a necessary precursor for zeolites in deep-sea sediments.

Depth of Burial

The majority of phillipsite and clinoptilolite occurrences are at depths less than
100 m and less than 600 m, respectively, below the sediment-water interface (Figure 3).
There is considerable overlap in the depth distribution of the two minerals. Considering
the relatively shallow depths involved and the relatively low temperatures, zeolites in
deep-sea sediments generally cannot be attributed to burial metamorphism. Even assuming
thermal gradients as high as 50°C/km and temperatures at the sediment-water interface of
4°C, the majority of phillipsite and clinoptilolite has formed at temperatures less than
10°C and 34°C, respectively.

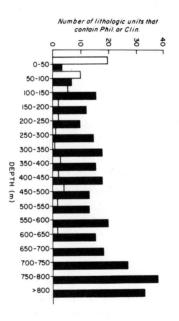

Figure 3. Distribution of deep-sea phillipsite and clinoptilolite below the sediment-
water interface (Stonecipher, 1978).

Areal Distribution

The majority of occurrences found in DSDP cores are in the Pacific and Indian Oceans although this distribution pattern may be biased by the large number of cores in these areas (Figure 4). Phillipsite is apparently much less common in the Atlantic than the Pacific, and Stonecipher (1976) attributed this to the fact that unlike the Atlantic, the Pacific Ocean has vast areas of sea floor below the calcite-compensation depth.

In surface sediments of the Pacific Basin, phillipsite is concentrated in zones north of 10°N latitude and south of 10°S latitude (Figure 5). The equatorial zone has generally low concentrations of this zeolite as does the East Pacific Rise. As noted by Cronan (1974), phillipsite distribution varies considerably on a local scale, even within regions where it is abundant. The same is true for clinoptilolite occurrences where closely spaced cores have been taken.

Associated Authigenic Minerals

Phillipsite is commonly associated with palagonite, smectite, iron and manganese oxides and/or hydroxides, and locally barite. Clinoptilolite is commonly associated with smectite, opal, cristobalite/tridymite (CT), chert, palygorskite, and sepiolite (see Kastner and Stonecipher, 1978). Clinoptilolite is more commonly associated with pyrite and siderite, whereas phillipsite is more commonly associated with barite. Clinoptilolite is more commonly associated with illite than is phillipsite (Stone- cipher, 1978), and some workers (Pim, 1973; von Rad and Rösch, 1972) believe that illite in older sediments may be diagenetic.

Mineralogy

Phillipsite. Phillipsite forms colorless to yellowish prismatic crystals from 8-250 μm long. The crystals form ball-shaped aggregates and complex-sector twins (Figure 6). The yellowish color in some crystals is due to minute isotropic inclusions of amorphous iron-bearing phases. Large phillipsites commonly show a zonal distribu- tion of the inclusions. The mean refractive index of the crystals is about 1.482 (Sheppard and Gude, 1970).

Chemical analyses of deep-sea phillipsites from the Pacific and Indian Oceans are given in Table 1. Potassium is usually the dominant exchangeable cation, and Si/Al ratios (2.4-2.8) are in an intermediate range for phillipsite-group minerals. Barium is generally present in minor amounts.

Table 1. Unit-cell compositions of deep-sea phillipsites (32 oxygen cell).[1]

	1	2	3	4	5	6	7	8	9	10	11	12	13	14	15
Si	11.65	11.71	11.79	11.58	11.46	11.61	11.38	11.56	11.39	11.38	11.78	11.40	11.54	11.92	11.50
Al	4.37	4.29	4.23	4.42	4.58	4.44	4.66	4.46	4.56	4.60	4.32	4.61	4.45	4.14	4.57
Mg	0.20	0.25	0.21	0.10	0.10	0.07	0.03	0.01	0.18	0.10	0.40	0.22	0.05	n.f.	n.f.
Ca	0.12	0.12	0.33	0.28	0.28	0.52	0.02	0.01	0.18	1.26	0.22	0.23	0.70	0.06	0.28
Ba	0.01	0.01	0.01	0.01	0.01	0.02	n.f.	n.f.	0.01	0.06	0.06	0.02	n.d.	n.d.	n.d.
Na	1.76	1.70	1.22	1.60	1.60	1.14	2.34	2.32	2.24	0.47	0.86	1.79	0.94	1.88	1.46
K	1.89	1.84	1.81	2.06	2.06	1.88	2.04	2.00	1.78	1.33	1.67	1.81	1.88	1.76	2.24
H_2O	10.31	9.91	10.99	11.37	11.01	11.11	12.31	10.92	11.85	12.76	11.04	1.027	n.d.	n.d.	n.d.
Si/Al	2.67	2.73	2.79	2.62	2.50	2.61	2.44	2.59	2.50	2.47	2.73	2.47	2.59	2.87	2.51

[1]Analyses 1 through 12 from Sheppard and Gude et al. (1970).
Analyses 13 through 15 from Stonecipher (1978).
n.f. = not found.
n.d. = not determined.

143

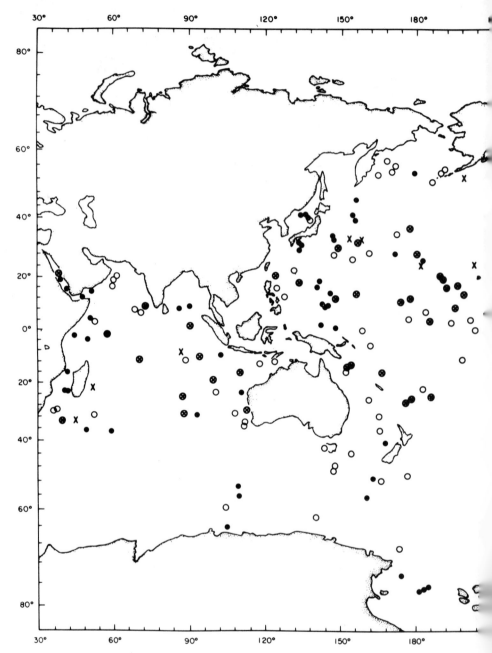

Figure 4. Distribution of phillipsite (X) and clinoptilolite (O) in subsurface samples of the world ocean basins. Solid dot = DSDP core site with no phillipsite or clinoptilolite found in bulk-rock x-ray diffraction analysis. See Figure 1a for data source.

144

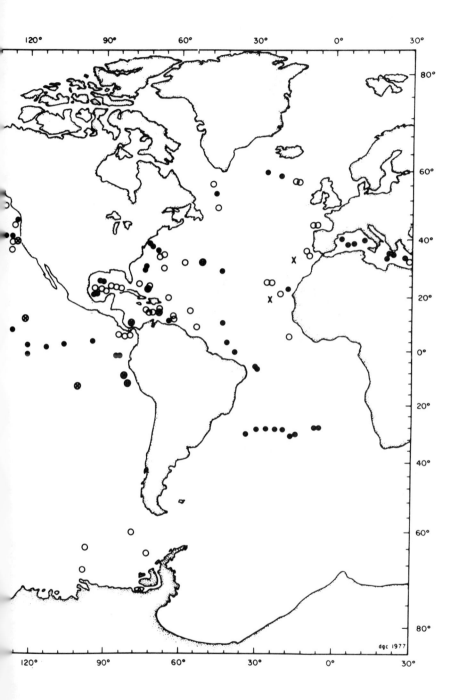

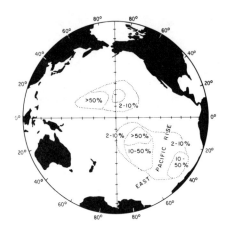

Figure 5.   Distribution of phillipsite in surface sediments of the Pacific Basin
(after Bonatti, 1963).

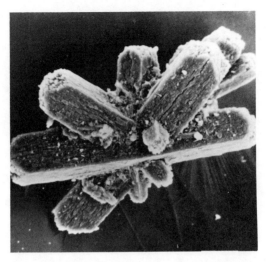

Figure 6. Phillipsite from radiolarian-rich clay. Scan 16 p, 200 cm, 16°25'N 164°24'W.
Bar scale = 25 μm. Photo courtesy of S. A. Stonecipher.

Attempts at K/Ar dating of phillipsites have been unsuccessful owing to the extensive growth of phillipsite after nucleation in the sediment column (Bernat et al., 1970).
Thorium and uranium contents of the phillipsites also decrease with depth in shallow
samples.

Clinoptilolite. Clinoptilolite forms clear, generally inclusion-free, prismatic
to barrel-shaped crystals commonly from 2 to 40 μm long (Figure 7). Mean refractive
indices range from 1.480-1.486.
Chemical analyses of deep-sea clinoptilolites are given in Table 2. Si/Al ratios
generally range from 4.1-4.9. Potassium is usually the dominant exchangeable cation.
Barium and strontium are not present in detectable amounts in the samples analyzed by
Boles and Wise (1978).
Clinoptilolites analyzed by Stonecipher (1978) have consistently higher Si/Al
ratios, generally lower sodium, and in some cases, significantly higher calcium contents
than samples analyzed by Boles and Wise (1978). These differences may be due more to
sample preparation, analytical techniques, and to the difficulty of probing fine-grained
crystals than to real differences. Samples analyzed by Boles and Wise were centrifuged
in distilled water to obtain a zeolite concentrate and analyzed by electron microprobe

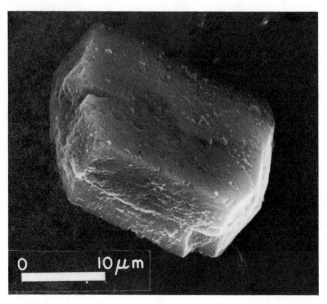

Figure 7a. Clinoptilolite (variety heulandite) from Paleocene lignite, DSDP site 214, Indian Ocean. Sample from 390.4 m below sediment-water interface.

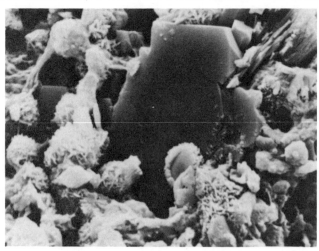

Figure 7b. Clinoptilolite from silicified limestone, DSDP site 116, north Atlantic Ocean. Sample from 850 m below sediment-water interface. Bar scale = 5 μm. Photo courtesy of S. A. Stonecipher.

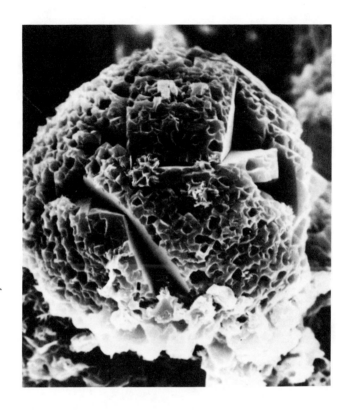

Figure 7c. Clinoptilolite in the form of radiolarian molds. DSDP site 9, Atlantic
Ocean. Core 3, section 3, 25 cm below top of section. Diameter of sphere
is approximately 200 μm. Photo courtesy of S. A. Stonecipher.

Table 2. Unit-Cell Compositions of Deep-Sea Clinoptilolites (72 oxygen cell)[1]

	1	2	3	4	5	6	7	8	9	10	11	12
Si	29.71	29.35	29.38	29.21	28.99	28.18	29.77	29.73	29.84	30.00	29.82	29.73
Al	6.37	6.62	6.88	6.88	7.05	7.89	6.28	6.19	6.15	6.06	6.12	6.23
Fe	0.06	0.06	0.04	0.17	0.02	0.01	n.f.	n.f.	n.f.	n.f.	n.f.	n.f.
Mg	0.39	0.38	n.f.	0.21	0.41	0.57	0.18	0.20	0.23	0.38	0.64	0.40
Ca	0.11	0.22	0.24	0.11	0.45	0.45	0.22	0.25	1.20	0.52	1.63	0.29
Na	2.32	2.19	2.19	2.69	2.13	2.46	1.75	1.98	0.65	1.03	0.37	1.06
K	2.58	3.16	3.07	2.67	2.96	3.06	3.53	3.45	2.68	2.98	1.45	3.58
Si/Al	4.66	4.43	4.27	4.25	4.11	3.57	4.74	4.80	4.85	4.95	4.87	4.77

[1] Analyses 1 through 6 from Boles and Wise (1978).
Analyses 7 through 12 from Stonecipher (1978).
n.f. = not found.
n.d. = not determined.

using techniques described by Boles (1972). Stonecipher (personal communication, 1977) acidized carbonate-rich samples to obtain a zeolite concentrate and treated the concentrate with sea water. Samples were analyzed with an energy dispersive system on an electron microprobe.

Analysis 6 of Table 2 is interesting because of its relatively low Si/Al ratio. This sample has an intermediate thermal stability with respect to most heulandite-clinoptilolites (see Boles and Wise, 1978) and is classified as a heulandite using the scheme of Boles (1972).

Figure 8 shows that deep-sea phillipsites and clinoptilolites have similar proportions of sodium, potassium, and calcium. Deep-sea clinoptilolites have high K/Na ratios relative to most clinoptilolites and heulandites from other settings.

Table 3 gives the compositions of interstitial waters associated with deep-sea phillipsite and clinoptilolite and suggests that the zeolites strongly fractionate potassium relative to sodium, calcium, and magnesium. The deep-sea clinoptilolite analyzed by Boles and Wise (1978) have K/Na/Ca/Mg ratios of 1/.80/.08/.10. Relative to potassium, sodium has been depleted by a factor of about 80, calcium by about 140, and magnesium by about 50 compared to their atomic proportions in the pore fluids (Boles and Wise, 1978).

The anomously low Si/Al ratio of sample 6 of Table 2 may have been caused by the presence of abundant organic material in the sediment at the time of formation, a situation which has been found to reduce silica solubility markedly (see analysis 3, Table 3; also Siever, 1962).

Origin of Phillipsite and Clinoptilolite

Palagonite, an alteration product of volcanic glass, has been recognized as a precursor for phillipsite (e.g., Murray and Renard, 1891; Bonatti, 1963; Nayudu, 1964; Rex, 1967). In many samples, however, there is no clear textural evidence for a precursor. Kastner (1976) reported that the alteration of basaltic glass to smectite preceded the formation of phillipsite. Phillipsites have also been recognized co-existing with clinoptilolite in andesitic detritus (Zemmels et al., 1975); thus, phillipsite may form from glasses somewhat richer in silica.

Weaver (1968), Berger and von Rad (1972), and Cook and Zemmels (1972) suggested that clinoptilolite may form directly from basaltic glass by the addition of silica (mainly biogenic) and water. However, at present there is no conclusive textural evidence which demonstrates a precursor phase for deep-sea clinoptilolite, aside from rare cases where clinoptilolite (?) has replaced siliceous radiolarian tests (Figure 7).

151

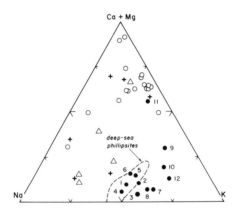

Figure 8. (Ca+Mg)-Na-K plot of the compositions of deep-sea clinoptilolites (solid dot) given in Table 1. Also plotted are the compositions of clinoptilolites from saline-lacustrine deposits (triangles) listed by Sheppard and Gude (1968, 1969, 1973); from marine tuff beds of New Zealand (open circles) listed by Boles (1972) and Boles and Coombs (1975); and from cavities in volcanic rocks (+) listed by Wise and Tschernich (1976) and from unpublished data of W. S. Wise. The field of deep-sea phillipsites includes all analyses from Sheppard and Gude (1970).

Table 3. Composition of Sea Water and of Interstitial Fluids (in g/kg) Associated with Zeolitic Sediments

| | Sediments with clinoptilolite | | | | | Sediments with phillipsite | |
	1	2	3	4	5	6	7
Na	10.6	10.7	10.6	9.9	9.8	10.60	10.79
K	0.38	0.32	0.24	0.27	0.22	0.47	0.47
Ca	0.40	1.19	1.34	0.71	1.57	0.46	0.32
Mg	1.27	0.80	0.82	1.16	1.00	1.27	1.32
Si	n.d.[8]	0.007	0.004	0.017	0.025	0.006	0.009
Cl	19.0	20.1	20.2	19.4	19.7	19.39	19.48
SO_4	2.75	1.90	1.64	2.30	2.24	2.02	2.31
HCO_3	0.14	n.d.	n.d.	0.07	0.05	0.09	--
pH	n.d.	7.8	6.7	7.0	6.7	7.5	7.5

1. "Average" sea water (Krauskopf, 1967).

2. Interstitial water from grayish-olive green silty sand, Paleocene, DSDP site 214, 364 m below sea bed (Manheim et al., 1974).

3. Interstitial water from grayish-black lignite, Paleocene(?), DSDP site 214, 400 m below sea bed (Manheim et al., 1974).

4. Interstitial water from zeolitic brown silty clay, Upper Cretaceous, DSDP site 137, 168 m below sea bed (Sayles and Manheim, 1975, Table 6).

5. Interstitial water from zeolitic clay, Upper Cretaceous, DSDP site 137, 225 m below sea bed (Sayles and Manheim, 1975, Table 6).

6. Interstitial water from brown zeolitic clay, undetermined age, DSDP site 51, 119 m below sea bed (Sayles and Manheim, 1975, Table 2).

7. Interstitial water from brown zeolitic clay, Miocene, DSDP site 50, 30 m below sea bed (Sayles and Manheim, 1975, Table 2).

8. n.d. = not determined.

Pseudomorphs after glass, such as those found in many other settings, have not been recognized. Clinoptilolite is not abundant in deep-sea vitric tuffs, but it is common in clay and/or carbonate-rich pelagic sediments in which evidence of volcanic detritus is sparse or absent. Many workers believe the presence of smectite, the clay commonly associated with clinoptilolite, indicates the former presence of volcanic glass (see Kastner and Stonecipher, 1977). Nevertheless, the absence of volcanic glass in most clinoptilolite-bearing sediments may indicate: (1) Clinoptilolite did not have a volcanic precursor; or (2) if it did, the volcanic glass has been completely dissolved. Perhaps in vitric ash beds the glass tends to be coarser grained and reaction rates are slow. The very low temperatures prevailing in deep-sea sediments would also serve to inhibit reaction rates; thus, only very fine-grained glass reacts completely.

Experimental work (see Stonecipher, 1976) indicates that the presence of carbonate catalyzes the conversion of smectite to illite and of biogenic silica to chert. By analogy, it is suggested that clinoptilolite crystallization may be catalyzed by the presence of carbonate, and thus, clinoptilolite is commonly associated with carbonate sediments. It should be noted that phillipsite is also commonly associated with carbonate sediments (Figure 2).

The spatial distribution of clinoptilolite and phillipsite in deep-sea sediments suggests a possible reaction relation between these zeolites (Berger and von Rad, 1972; Couture, 1976; Boles and Wise, 1978). The similar exchangeable cation content of phillipsite and clinoptilolite (Figure 8) suggests that such a reaction would require addition of only silica and water. The reaction might be:

phillipsite + quartz (or biogenic silica) + water = clinoptilolite.

Boles and Wise (1978) calculated about 10 cc of phillipsite would require about 3 cc quartz and 1 cc water to form 14-15 cc of clinoptilolite. If this reaction occurs, high silica activity would favor clinoptilolite over phillipsite; however, as shown in Table 3, the silica concentrations of pore fluids associated with these sediments are very similar. It is possible that the dissolution and recrystallization of biogenic silica to cristobalite or quartz releases silica for this reaction. Another possible source of silica is that released by the reaction of smectite to illite (cf. Stonecipher, 1976), although smectite is the dominant clay type associated with clinoptilolite.

The above considerations suggest that phillipsite forms metastably, probably as a silica-deficient phase with respect to marine pore fluids. Phillipsite can be readily crystallized from volcanic glass in the laboratory (Mariner and Surdam, 1970). It has been observed in a number of Pleistocene or younger, nonmarine, saline-alkaline

154

deposits (Hay, 1966, Table 4), but it is rare in such deposits of Eocene and older ages. Sheppard and Gude (1969, 1973) have suggested that phillipsite forms before clinoptilolite in saline-alkaline lake deposits. Their data imply that phillipsite is a readily formed, early phase which does not persist.

Although field and experimental evidence indicates that phillipsite is a common alteration product of volcanic glass, the fact that the most clinoptilolite-phillipsite occurrences are not associated with highly volcanogenic sediments (see Figure 2) is surprising. This may suggest that volcanic glass is not a prerequisite for phillipsite and clinoptilolite. Arrhenius (1963), Berger and von Rad (1972), and Boles and Wise (1978) suggested that dissolution of siliceous microfossils may provide a source of silica in the absence of glass. Hurd (1973) has shown that acid-cleaned opal tests can contain almost 0.77 weight % Al, which would presumably be released and utilized for zeolite formation upon dissolution of the tests. Clay minerals (e.g., smectites) may also be involved in such a reaction, possibly providing Al and other cations for zeolitization.

OTHER ZEOLITES IN DEEP-SEA SEDIMENTS

Analcime

Analcime was first reported in deep-sea sediments by Murray and Renard (1891). It has since been reported from at least 37 bulk-rock x-ray diffraction analyses of DSDP core samples. Most of the occurrences are in the Pacific and Indian Oceans. Stonecipher (1978) reported that improved x-ray diffraction techniques applied after drill site 249 have increased the number of reported occurrences; thus, analcime may be more common in deep-sea sediments than presently believed.

The number of analcime occurrences generally increases with sediment age in a similar way to clinoptilolite occurrences (Figure 9). The zeolite occurs in a variety of sediment types but is most common in calcareous oozes and volcanoclastic sediments (Figure 10). It commonly makes up only a few percent of the bulk sediment, but in early Cretaceous volcanogenic sandstones and siltstones from the Central Pacific (DSDP Site 317A), it makes up more than 25% of the bulk sediment. Analcime commonly coexists with clinoptilolite and locally with phillipsite. Smectite is the dominant clay type associated with analcime.

There are no chemical analyses of analcimes from deep-sea sediments, and there has been no systematic study of its origin in these sediments. Analcime may have been derived from a zeolite precursor such as clinoptilolite. This reaction is known to occur

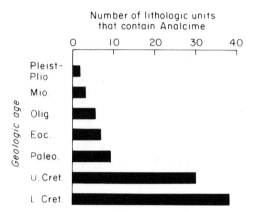

Figure 9. Distribution of deep-sea analcime with respect to age of sediment (modified after Stonecipher, 1978). Data have been normalized by Stonecipher to account for the coring frequency of a given sediment age.

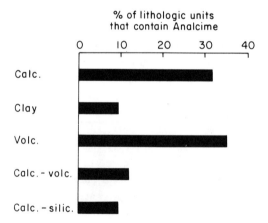

Figure 10. Distribution of deep-sea analcime with respect to sediment type (after Stonecipher, 1978). Calc = carbonate oozes with less than 10% volcanic ash; clay = clays and mudstone; volc = ash layers and volcanoclastic sandstones and mudstones; calc-volc = calcareous oozes interbedded or mixed with volcanic debris; calc-silic = radiolarian/nanno oozes or limestone with chert.

in saline alkaline-lake deposits and in marine sediments from burial metamorphic terraines. The reaction of deep-sea clinoptilolite to analcime, using the average clinoptilolite composition given by Boles and Wise (1978) and an analcime composition of
$NaAlSi_3O_3 \cdot H_2O$ would be:

$$4.4Na^+ + clinoptilolite \rightarrow 6.8 \text{ analcime} + 3K^+ + 0.3Ca^{+2}$$
$$+ 0.4Mg^{+2} + 15.6SiO_2 + 13.2H_2O.$$

Aside from pressure and temperature controls, such a reaction would be dependent mainly
on sodium, potassium, silica, and water activities. There are only scant data on pore-
fluid chemistry from analcime-bearing sediments, but some studies show that such pore
fluids have molar Na/K ratios of about 160 (Gieskes, 1976, Table 1, water analyses from
Site 317A). Molar Na/K ratios of pore fluids in clinoptilolite-bearing sediments
average about 60 (Table 3). Thus, the presence of analcime may be due to high Na/K
ratios in pore fluids; alternatively, the high Na/K ratios of the pore fluids may
simply reflect the presence of analcime.

Harmotone

Harmotone, the barium-rich member of the phillipsite group, has been reported to
coexist with phillipsite in deep-sea sediments (Murray and Renard, 1891; Goldberg and
Arrhenius, 1958; Arrhenius, 1963). Most of these occurrences are in pelagic clays from
the Pacific basin. Harmotone and phillipsite are also reported as a cement in Pleistocene scorias and tuffaceous mudstones from west of the Society Ridge, South Pacific
(Morgenstein, 1967).

Harmotone has been reported to be common in deep-sea sediments (Arrhenius, 1963),
but because harmotone and phillipsite have very similar x-ray diffraction patterns
(see Černy et al., 1977), it is difficult to estimate the overall abundance of this
phase relative to phillipsite in the absence of chemical data. Sheppard et al. (1970)
found less than 1 weight % BaO in the phillipsites they analyzed from the Indian and
Pacific Oceans (also see Goodell et al., 1970). Goldberg and Arrhenius (1958) reported
3-15 weight % BaO in "phillipsite" from Pacific pelagic sediments; this appears to be
the only chemically documented harmotone in deep-sea sediments. Thus, barium-rich
members of the group may not be common in deep-sea sediments (Cronan, 1974). Harmotone presumably has a similar origin as phillipsite, for example, as an alteration
product of palagonite (Morgenstein, 1967).

Laumontite

Laumontite has been identified only at DSDP drill site 323 in the southern Pacific Ocean (Sands and Drever, 1978). The mineral was identified at 162 m below the sediment-water interface in a diatom ooze, at 322 m in a sandstone, at 367 m in a diatom ooze, and at 697 m in a clinoptilolite-bearing clay-rich altered ash. The dominant clay type in the samples is smectite.

A semiquantitative chemical analysis of the laumontite (obtained by x-ray energy spectrometry) indicates SiO_2-48.3 wt. %, Al_2O_3-17.3 wt. %, Fe_2O_3-4.4%, CaO-9.2%, and K_2O-4.1%. This laumontite is considerably richer in iron and potassium than laumontite from low-grade metamorphic terraines.

Sands and Drever (1978) suggested that laumontite formation was favored by high calcium and low silica activities in the interstitial waters (Gieskes and Lawrence, 1976). Horne (1968) reported laumontite and epidote in Cretaceous sediments of Antarctica but Sands and Drever concluded that the laumontite is not detrital, based on the absence of epidote (compare Kastner and Stonecipher, 1978). If this laumontite is authigenic, it must have formed at temperatures significantly less than presently believed. Laumontite may therefore not be a reliable indicator of very low-grade metamorphic conditions.

Erionite

Erionite has been identified in three samples at DSDP drill site 60 in the western Pacific (Rex et al., 1971). The mineral occurs at about 350 m below the sediment-water interface in early Miocene altered ash in quantities exceeding 20% of the bulk sample.

Zeolites in Volcanic Rocks

Various zeolites have been found as alteration of basaltic rocks which underlie deep-sea sediments. For example, Bass et al. (1973) reported phillipsite, chabazite, gmelinite, natrolite, and thomsonite as replacement products or veins in basaltic rocks of the South Atlantic. These types of zeolite occurrences have not been studied in detail and will not be reviewed here.

COMPARISON OF DEEP-SEA ZEOLITE
OCCURRENCES WITH OTHER OCCURRENCES

Conditions of Formation

Deep-sea zeolites have generally formed at temperatures slightly to significantly less than in other environments. Most of the zeolites have probably formed at temperatures less than 10°-34°C. Higher temperatures may have prevailed in some areas for relatively short time spans due to low-temperature hydrothermal activity. Fluid pressures in which these zeolites have commonly formed are less than about 0.7 kb, judging from their depth of occurrence below the sediment-water interface and assuming maximum water depths of 6000 m.

In general, the interstitial fluids in zeolitic deep-sea sediments are similar to sea water, and marked deviations in pH and other cations are generally not found. In contrast, zeolites in saline alkaline-lakes and, perhaps, in low-grade metamorphic settings have formed in association with fluids of widely varying compositions.

Distribution of Zeolites

Phillipsite, clinoptilolite, and, to a lesser extent, analcime are the dominant zeolites in deep-sea sediments. In this regard, deep-sea zeolite assemblages are much simpler than those in other settings, reflecting the uniformity in conditions at which they formed. Phillipsite is more characteristic of Miocene and younger sediments, whereas clinoptilolite and analcime characterize older sediments. Hay (1966) noted a similar relationship for on-land zeolite occurrences. These observations suggest that reaction kinetics are an important factor in controlling the presence of a given zeolite and that a phase such as phillipsite forms metastably (compare Glaccum and Boström, 1976).

Mode of Zeolite Occurrences

Zeolites in deep-sea sediments occur in widely varying lithologies but are most common in very fine-grained pelagic sediments where they occur as euhedral crystals in the matrix. In most occurrences volcanogenic detritus is not obvious. Zeolite occurrences in other settings are, in most cases, associated with highly volcanogenic sediments, and pseudomorphs[1] after volcanic glass are very common. Whether or not these

----

[1]Pseudomorph is used in the petrographic sense in that clusters of individual crystals mimic the shapes of pre-existing glass shards.

differences reflect a different mode of origin for deep-sea occurrences versus other types of zeolite occurrences is uncertain. Certainly there is reason to doubt that all zeolite occurrences in deep-sea sediments have originated from the direct alteration of volcanic glass. In part, the differences in mode of occurrences may reflect differences in reaction temperatures. Perhaps only the finest grained ash associated with pelagic sediments can react in appreciable quantities to form zeolites at the prevailing low temperatures.

REFERENCES

Arrhenius, G. (1963) Pelagic sediments: In, Hill, M.N., Ed., The Sea, Vol. 3, Wiley Interscience, N.Y., 655-727.

Bass, M.N. (1976) Secondary minerals in oceanic basalts, with special reference to Leg 34, Deep Sea Drilling Project: Yeats, R.S. et al., Eds., Initial Reports of the Deep Sea Drilling Project, Vol. XXXIV, U.S. Gov. Printing Office, Washington, D.C., 393-432.

Bass, M.N., Moberly, R., Rhodes, J.M., Shih, Chi-yu, and Church, S.E. (1973) Volcanic rocks cored in the Central Pacific, Leg 17, Deep Sea Drilling Project: Winterer, E.L. et al., Eds., Initial Reports of the Deep Sea Drilling Project, Vol. XVII, U.S. Gov. Printing Office, Washington, D.C., 429-503.

Benson, R.N. (1972) Radiolaria, Leg 12, Deep Sea Drilling Project: Ed., Initial Reports of the Deep Sea Drilling Project, Vol. XII, U.S. Gov. Printing Office, Washington, D.C., 1085-1113.

Berger, W.H. and von Rad, U. (1972) Cretaceous and Cenozoic sediments from the Atlantic Ocean: Hayes, D.F. et al., Eds., Initial Reports of the Deep Sea Drilling Project, Vol. XIV, U.S. Gov. Printing Office, Washington, D.C., 787-954.

Bernat, M., Bieri, R.H., Koide, M., Griffin, J.J., and Goldberg, E.O. (1970) Uranium, thorium, potassium and argon in marine phillipsites: Geochim. Cosmochim. Acta 34, 1053-1071.

Biscaye, P.E. (1965) Mineralogy and sedimentation of Recent deep-sea clay in the Atlantic Ocean and adjacent seas and oceans: Geol. Soc. Am. Bull. 76, 803-832.

Boles, J.R. (1972) Composition, optical properties, cell dimensions, and thermal stability of some heulandite group zeolites: Am. Mineral. 57, 1463-1493.

Boles, J.R. and Coombs, D.S. (1975) Mineral reactions in zeolitic Triassic tuffs, Hokonui Hills, New Zealand: Geol. Soc. Am. Bull. 86, 163-173.

Boles, J.R. and Wise, W.S. (1978) Nature and origin of deep-sea clinoptilolite: In, Sand, L.B. and Mumpton, F.A., Eds., Natural Zeolites: Occurrence, Properties, Use, Pergamon Press, Elmsford, N.Y., 235-243.

Bonatti, E. (1963) Zeolites in Pacific pelagic sediments: Trans. N.Y. Acad. Sci. 25, 938-948.

Bonatii, E. and Nayudu, Y.R. (1965) The origin of manganese nodules on the ocean floor: Am. J. Sci. 263, 17-39.

Cerný, P., Rinaldi, R., and Surdam, R.C. (1977) Wellsite and its status in the phillip-site-harmotone group: Neues Jahrb. Mineral. Abh. 128, 312-330.

Cook, H.E. and Zemmels, I. (1972) X-ray mineralogy studies--Leg 9: In, Hayes, J.D. et al., Eds., Initial Reports of the Deep Sea Drilling Project, Vol. IX, U.S. Gov. Printing Office, Washington, D.C., 707-778.

Couture, R.A. (1976) Composition and origin of palygorskite-rich and montmorillonite-rich zeolite-containing sediments from the Pacific Ocean: Chem. Geol. 13, 113-130.

Cronan, D.S. (1974) Authigenic minerals in deep-sea sediments: In, Goldberg, E.D., Ed., The Sea, Vol. 5, Wiley Interscience, N.Y., 491-525.

Czyscinski, K. (1973) Authigenic phillipsite formation rates in the Central Indian Ocean and the Euqatorial Pacific Ocean: Deep-Sea Res. 20, 555-559.

Gieskes, J.M. (1976) Interstitial water studies, Leg 33: In, Schlanger, S.O. et al., Eds., Initial Reports of the Deep Sea Drilling Project, Vol. XXXIII, U.S. Gov. Printing Office, Washington, D.C., 563-570.

Gieskes, J.M. and Lawrence, P.R. (1976) Interstitial water studies, Leg 35, In, Hollister, C.D. et al., Eds., Initial Reports of the Deep Sea Drilling Project, Vol. XXXV, U.S. Gov. Printing Office, Washington, D.C., 407-424.

Glaccum, R. and Boström, K. (1976) (Na-K)-phillipsite: Its stability conditions and geochemical role in the deep sea: Marine Geol. 21, 47-58.

Goldberg, E.D. and Arrhenius, G. (1958) Chemistry of Pacific pelagic sediments: Geochim. Cosmochim. Acta 13, 153-212.

Goodell, H.G., Meylan, M.A., and Grant, B. (1970) Ferromanganese deposits of the South Pacific Ocean, Drake Passage and Scotia Sea: In, Reid, J.L., Ed., Antarctic Oceanology, Vol. 1, American Geophysical Union, Washington, D.C., 27-92.

Hathaway, J.C. and Sachs, P.L. (1965) Sepiolite and clinoptilolite from the mid-Atlantic ridge: Am. Mineral. 50, 852-867.

Hay, R.L. (1966) Zeolites and zeolitic reactions in sedimentary rocks: Geol. Soc. Am. Spec. Pap. 85, 130 pp.

Heezen, B.C. and Hollister, C.D. (1971) The Face of the Deep: Oxford Press, New York, N.Y., 659 pp.

Horne, R.R. (1968) Authigenic prehnite, laumontite and chlorite in the Lower Cretaceous sediments of south-eastern Alexander Island: Br. Antarct. Surv. Bull. 18, 1-10.

Hurd, D.C. (1973) Interactions of biogenic opal, sediment and sea water in the Central Equatorial Pacific: Geochim. Cosmochim. Acta 37, 2257-2282.

Kastner, M. (1976) Diagenesis of basal sediments and basalts of sites 322 and 323, Leg 35, Bellingshausen Abyssal Plain: In, Hollister, C. and Cradock, C., Eds., Initial Reports of the Deep Sea Drilling Project, Vol. XXXV, U.S. Gov. Printing Office, Washington, D.C., 513-528.

Kastner, M. and Stonecipher, S.A. (1978) Zeolites in deep-sea sediments: In, Sand, L.B. and Mumpton, F.A., Eds., Natural Zeolites: Occurrence, Properties, Use, Pergamon Press, Elmsford, N.Y., 199-220.

Kolla, V. and Biscaye, P.E. (1973) Deep-sea zeolites: variations in space and time in sediments of the Indian Ocean: Marine Geol. 15, 11-17.

Krauskopf, K. (1967) Introduction to Geochemistry, McGraw-Hill Publ. Co., N.Y., 721 pp.

Manheim, F.T., Waterman, L.S., and Sayles, F.L. (1974) Interstitial water studies on small cores, Leg 22: In, Initial Reports of the Deep Drilling Project, Vol. XXII, U.S. Gov. Printing Office, Washington, D.C., 657-662.

Mariner, R.H. and Surdam, R.C. (1970) Alkalinity and formation of zeolites in saline-alkaline lakes: Science 170, 977-979.

Morgenstein (1967) Authigenic cementation of scoriaceous deep-sea sediments west of the Society Ridge, South Pacific: Sedimentol. 9, 105-118.

Murray, J. and Renard, A.F. (1891) Report on deep-sea deposits: Report on the Scientific Results of the Voyage of the H.M.S. Challenger During the Years 1873-76, Neill and Co., Edinburgh, 520 pp.

Nayadu, Y.R. (1964) Palagonite tuffs (hyaloclastites) and the products of post-eruptive processes: Bull. Volcanol. 27, 391-410.

Pim, A.C. (1973) Trace element determinations compared with x-ray diffraction results of brown clay in the Central Pacific: In, Winterer, E.L. et al., Eds., Initial Reports of the Deep Sea Drilling Project, Vol. XVII, U.S. Gov. Printing Office, Washington, D.C., 511-514.

Rex, R.W. (1967) Authigenic silicates formed from basaltic glass by more than 60 million years' contact with sea water: Clays and Clay Minerals 15, 195-203.

Rex, R.W., Eklund, W.A., and Jamieson, I.M. (1971) X-ray mineralogic studies, Leg 6: In, Fischer, A.G. et al., Eds., Initial Reports of the Deep Sea Drilling Project, Vol. VI, U.S. Gov. Printing Office, Washington, D.C., 753-810.

Robinson, P.T., Thayer, P.A., Cook, P.J., and McKnight, B.T. (1974) Lithology of Mesozoic and Cenozoic sediments from the Indian Ocean, Leg 27: In, Veevers, J.J. et al., Eds., Initial Reports of the Deep Sea Drilling Project, Vol. XXVII, U.S. Gov. Printing Office, Washington, D.C., 1001-1048.

Sands, C.D. and Drever, J.I. (1978) Authigenic laumontite in deep sea sediments: In, Sand, L.B. and Mumpton, F.A., Eds., Natural Zeolites: Occurrence, Properties, Use, Pergamon Press, Elmsford, N.Y., 269-275.

Sayles, F.L. and Manheim, F.T. (1975) Interstitial solutions and diagenesis in deeply buried marine sediments: results from the Deep Sea Drilling Project: Geochim. Cosmochim. Acta 39, 103-127.

Siever, R. (1962) Silica solubility, 0°-200°C, and the diagenesis of siliceous sediments: J. Geol. 70, 127-150.

Sheppard, R.A. and Gude, A.J., 3rd (1968) Distribution and genesis of authigenic silicate minerals in tuffs of Pleistocene Lake Tecopa, Inyo County, California: U.S. Geol. Surv. Prof. Pap. 597, 38 pp.

Sheppard, R.A. and Gude, A.J., 3rd (1969) Diagenesis of tuffs in the Barstow Formation, Mud Hills, San Bernardino County, California: U.S. Geol. Surv. Prof. Pap. 634, 35 pp.

Sheppard, R.A., Gude, A.J., 3rd, and Griffin, J.J. (1970) Chemical composition and physical properties of phillipsite from the Pacific and Indian Oceans: Am. Mineral. 55, 2053-2062.

Sheppard, R.A. and Gude, A.J., 3rd (1973) Zeolites and associated minerals in tuffaceous rocks of the Big Sandy Formation, Mohave County, Arizona: U.S. Geol. Surv. Prof. Pap. 830, 36 pp.

Stonecipher, S.A. (1976) Origin, distribution and diagenesis of phillipsite and clinoptilolite in deep-sea sediments: Chem. Geol. 17, 307-318.

Stonecipher, S.A. (1978) Chemistry of deep-sea phillipsite, clinoptilolite, and host sediment: In, Sand, L.B. and Mumpton, F.A., Eds., Natural Zeolites: Occurrence, Properties, Use, Pergamon Press, Elmsford, N.Y., 221-234.

von Rad, U. and Rösch, H. (1972) Mineralogy and origin of clay minerals, silica and authigenic silicates in Leg 14 sediments: In, Hayes, D.E. et al., Eds., Initial Reports of the Deep Sea Drilling Project, Vol. XIV, U.S. Gov. Printing Office, Washington, D.C., 727-739.

Weaver, C.E. (1968) Mineral facies of the Tertiary of the Continental shelf and Blake Plateau: Southeastern Geol. 9, 57-63.

Wise, W.S. and Tschernich, R.W. (1976) Chemical composition of ferrierite: Am. Mineral. 61, 60-67.

Zemmels, I., Cook, H.E., and Matti, J.C. (1975) X-ray mineralogy data, Tasman Sea and far Western Pacific, Leg 30, Deep Sea Drilling Project: In, Andrews, J.E. et al., Initial Reports of the Deep Sea Drilling Project, Vol. XXX, U.S. Gov. Printing Office, Washington, D.C., 603-616.

Chapter 8

COMMERCIAL PROPERTIES OF NATURAL ZEOLITES

E. M. Flanigen and F. A. Mumpton

INTRODUCTION

All commercial applications of natural zeolites make use of one or more of several
physical or chemical properties, including (1) ion exchange, (2) adsorption and related molecular sieve properties, (3) dehydration and rehydration, and (4) siliceous
composition. These properties, of course, are functions of the specific crystal structure of each individual zeolite species and of their framework and cationic compositions.
In addition, certain extrinsic properties, such as the tendency of "sedimentary" zeolites
to occur as light-colored, lightweight, porous aggregates of micrometer-size crystals,
have contributed to their past and present-day uses.

This chapter covers in abbreviated form several of the important chemical and
physical properties of zeolite materials which are currently being exploited in industrial and agricultural technology, including cation exchange, adsorption, dehydration,
and thermal stability.

ADSORPTION PROPERTIES

Crystalline zeolites are unique adsorbent materials. Under normal conditions, the
large central cavities and entry channels of zeolites are filled with water molecules
forming hydration spheres around the exchangeable cations. If the water is removed,
usually by heating to 350° or 400°C for a few hours or overnight, molecules having
effective cross-sectional diameters small enough to pass through the entry channels
are readily adsorbed in the dehydrated channels and central cavities. Molecules too
large to pass through the entry channels are excluded, giving rise to the well-known
"molecular sieving" property of most zeolites (see Figure 1). By way of illustration,
consider the calcium-exchanged version of synthetic zeolite A which has pores about
4.5 Å in diameter. Normal hydrocarbons, such as pentane and octane, having effective

165

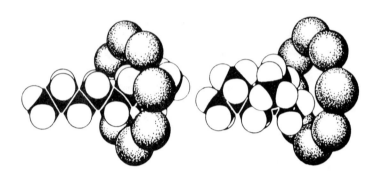

Figure 1. Schematic illustration of the entry of straight-chain hydrocarbons and the
blockage of branch-chain hydrocarbons at channel apertures.

cross-sectional diameters of about 4.3 Å are easily taken up by this zeolite; however,
branch-chain hydrocarbons, such as iso-pentane and iso-octane, with diameters of 5.0 Å
or larger are essentially not adsorbed by this material. Thus, the separation of mixtures of straight- and branch-chain hydrocarbons and of paraffin and aromatic hydrocarbons can be accomplished by passing the gaseous stream through columns packed with
dehydrated zeolites selected for their pore-size distribution.

Because of the uniform size of the rings of oxygens in their framework structures,
zeolites have relatively narrow pore-size distributions, in contrast to the wide range
of pore sizes of other commercial adsorbents, such as silica gel, activated alumina, and
activated carbon (Figure 2). Adsorption on crystalline zeolites is therefore characterized by Langmuir-type isotherms, such as that shown in Figure 3. Note that the
quantity adsorbed (x), relative to the quantity of complete pore-filling (x_s), is
maximized at very low partial pressures of the adsorbate. While the total amount of a
gas adsorbed may not be as great as for other types of adsorbents, crystalline zeolites
are excellent adsorbents for removing the last trace of a particular gas from a system.
This property is especially important in certain dessication applications where it is
essential to lower the water content to less than 0.1 ppm.

The surface area available for adsorption ranges up to several hundred square
meters per gram, and some zeolites are capable of adsorbing up to about 30% of their
dry weight. Most of the surface area is found within the zeolite structure and

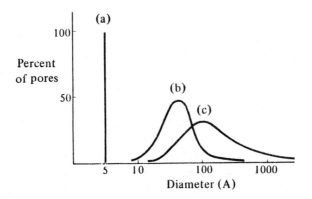

Figure 2.   Distribution of pore sizes in microporous adsorbents. (a) Dehydrated
crystalline zeolite; (b) typical silica gel; (c) activated carbon.
(From Breck, 1974, Fig. 1.1.)

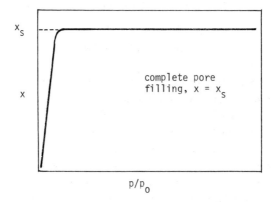

Figure 3.   Langmuir-type isotherm for adsorption on crystalline zeolites illustrating
almost complete saturation at low partial pressures of the adsorbate. x =
amount adsorbed; p = pressure.

167

represents the inner surface of dehydrated channels and cavities. Only about 1% is contributed by the external surface of the zeolite particle.

In addition to their ability to separate gas molecules on the basis of size and shape, the unusual charge distribution within the dehydrated void volume due to the presence of cations, hydroxyl groups, and field gradients generated by the substitution of aluminum for silicon in the framework, allows many species with permanent dipole moments to be adsorbed with a selectivity unlike that of almost any other adsorbent. Thus, polar molecules such as H_2O, CO_2, and H_2S are adsorbed preferentially over non-polar molecules, and adsorption processes have been developed using natural zeolites to remove CO_2 and other contaminants from impure natural gas and other methane streams to give almost pure CH_4 products. In addition, the small but finite, quadripole moment of N_2 allows it to be adsorbed selectively from air, thereby providing a means of producing oxygen-enriched streams at room temperature by pressure-swing adsorption on certain natural or synthetic zeolites.

Many zeolites can therefore act as selective adsorbents or "molecular sieves," separating gaseous molecules on the basis of size, shape, and surface selectivity. In addition to the many synthetic zeolites used commercially in this manner, certain abundant natural species, such as erionite, chabazite, mordenite, and clinoptilolite, have potential in these areas also, as discussed in Chapter 9.

CATION-EXCHANGE PROPERTIES

The cation-exchange properties of zeolite minerals were first observed more than a century ago (Eichorn, 1858) and have recently been the subject of extensive reviews by Barrer (1974, 1978) and Sherry (1971). The exchangeable cations of a zeolite are only loosely bonded to the tetrahedral framework and can be removed or exchanged easily by washing with a strong solution of another ion. As such, crystalline zeolites are some of the most effective ion exchangers known to man, with capacities of up to 3 or 4 milliequivalents per gram. This compares favorably with most montmorillonitic clay minerals which have exchange capacities of about 0.8-1.0 meq/g. The ion-exchange capacity is basically a function of the degree of substitution of aluminum for silicon in the framework structure; the greater the substitution, the greater the charge deficiency, and the greater the number of alkali or alkaline earth cations required for electrical neutrality. In practice, however, the cation-exchange behavior is dependent on a number of other factors as well, including (1) the nature of the cation species (size, charge, etc.), (2) temperature, (3) concentration of the cation species in solution, and (4) the structural characteristics of the particular zeolite under investigation.

In certain species, cations may be trapped in structural positions that are relatively inaccessible, thereby reducing the effective exchange capacity of that species for that ion. Also, cation sieving may take place if the size of the cation in solution is too large to pass through entry ports into the central cavities of the structure. Analcime, for example, will exchange almost completely its Na^+ for Rb^+ (ionic radius = 1.49 Å), but not at all for Cs^+ (ionic radius = 1.65 Å) (Breck, 1974).

Unlike most non-crystalline ion exchangers, such as organic resins or inorganic aluminosilicate gels (mislabeled in the trade as "zeolites"), the framework of a crystalline zeolite dictates its selectivity towards competing ions, and different structures offer different sites for the same cation. The hydration spheres of high field-strength ions prevent their close approach to the seat of charge in the framework; therefore, in many zeolites, ions with low field strength are more tightly held and selectively taken up from solution than other ions. For example, in the zeolite clinoptilolite, the small amount of aluminum substituting for silicon in the framework results in a relatively low ion-exchange capacity (about 2.3 meq/g); however, its cation selectivity is as follows:

$$Cs > Rb > K > NH_4 > Ba > Sr > Na > Ca > Fe > Al > Mg > Li \quad (Ames, 1960).$$

Thus, clinoptilolite has a decided preference for larger cations as does the zeolite chabazite which has the following selectivity:

$$Tl > K > Ag > Rb > NH_4 > Pb > Na = Ba > Sr > Ca > Li \quad (Breck, 1974, p. 556).$$

Synthetic zeolite A, on the other hand, shows a widely different type of cation selectivity, as evidenced by the following sequences for mono- and divalent cations:

$$Ag > Tl > Na > K > NH_4 > Rb > Li > Ca$$

$$Zn > Sr > Ba > Ca > Co > Ni > Cd > Hg > Mg$$

(Breck, 1974, p. 538). Type A zeolite is also more selective for Ca than for Na, explaining its recent introduction as a water softener in phosphate-free detergents. The preference for larger cations, such as NH_4^+, was exploited by Ames (1967) and Mercer et al. (1970) in the development of an ion-exchange process for the removal of ammoniacal nitrogen from municipal and agricultural wastewaters. Clinoptilolite, mordenite, and chabazite have also been used in the treatment of radioactive wastes because of their affinity for Cs and Sr, as discussed in Chapter 9.

The cation-exchange reaction can be expressed simply as:

$$M^1(Z) + M^2(S) = M^2(Z) + M^1(S)$$

where M^1 is the cation in the zeolite (Z), and M^2 is the cation in the solution (S). The solvent is typically water, but it may be any solvent in which the salt of M^2 is soluble. Ion exchange between a zeolite and a solid salt has only been reported at elevated temperatures. Cation-exchange equilibria between a zeolite and a solution are usually depicted by an ion-exchange isotherm which plots equivalent molal fraction of the exchanging cation in the zeolite phase (A_Z) as a function of the equivalent molal fraction of the exchanging cation in the solution phase (A_S). The different kinds of selectivities and isotherm shapes shown in Figure 4 reflect the diversity of zeolite frameworks and the stabilities of cations in various sites within the structures.

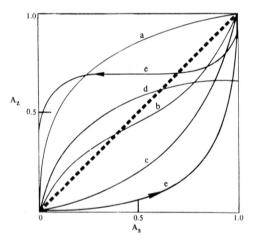

Figure 4. Types of ion-exchange isotherms for the reaction $A_S + B_Z = A_Z + B_S$. Five types of isotherms are illustrated: (a) selectivity for the entering cation over the entire range of zeolite composition; (b) the entering cations shows a selectivity reversal with increasing equivalent fraction in the zeolite; (c) selectivity for the leaving cation over the entire range of zeolite composition; (d) exchange does not go to completion although the entering cation is initially preferred (the degree of exchange, $x_{max} < 1$ where x is the ratio of equivalents of entering cation to the gram. equiv. of Al in the zeolite); (e) hysteresis effects may result from formation of two zeolite phases. (From Breck, 1974, Fig. 7.2.)

DEHYDRATION AND DEHYDROXYLATION PHENOMENA

Based on dehydration behavior, zeolites may be classified as (a) those which show
no major structural changes during dehydration and which exhibit continuous weight-loss
curves as a function of temperature, and (b) those which undergo major structural
changes during dehydration and which exhibit discontinuities in their weight-loss curves.
The latter group includes those zeolites whose structures collapse on heating to elevated
temperature. Figure 5a and 5b show schematic differential thermal analysis (DTA) and
thermal gravimetric analysis (TGA) curves for zeolites of the first type. These curves
are characteristic or rigid, three-dimensional zeolite structures, such as synthetic
zeolites A and X, natural chabazite, mordenite, erionite, and clinoptilolite which are
thermally stable to 700° or 800°C. The dehydration behavior of zeolites of the second
type is illustrated by the DTA and TGA curves for the natrolite minerals, as shown in
Figure 6. In these materials, water molecules are arranged in definite groups within
the structure, each group having a different volatility. In general, such materials
exhibit typical reversible water loss at low temperatures; however, once a major structural change occurs, as evidenced by a sharp endotherm and an abrupt change of slope of
the weight-loss curve, the material loses its zeolitic character. A summary of the dehydration behavior of zeolites is listed in Table 6.1 of Breck (1974).

Structural hydroxyl groups can be introduced into zeolites by (1) cation hydrolysis
or (2) deammoniation of NH_4-exchanged zeolites. Normal dehydration causes dissociation
of water molecules by the electrostatic field of the association cation:

$$Ca^{2+}(H_2O)_x \quad \text{(structure)} \longrightarrow \quad H \text{ (structure)} + CaOH^+$$

Thermal decomposition of ammonium-exchanged zeolites follows the reaction scheme:

$$NH_4^+ \quad NH_4^+ \quad \text{(structure)} \xrightarrow{-NH_3} \quad H \quad H \text{ (structure)}$$

This reaction is also referred to as "decationization." The presence of hydroxyl groups
in zeolites is especially important in their use as catalysts. The active sites for
hydrocarbon conversions and other catalytic reactions are acidic protons associated with
structural OH groups. The reaction to form hydroxyl groups in zeolites described above
occur below 500°C. At higher temperatures, dehydroxylation occurs according to the
following idealized scheme:

171

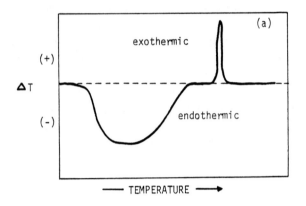

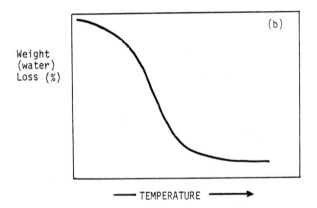

Figure 5.   (a) Schematic differential thermal analysis (DTA) curve for a zeolite undergoing no major structural change on dehydration. Note the broad, low-temperature endotherm caused by loss of water and the high-temperature exotherm
due to structural transformation. (b) Thermal gravimetric analysis (TGA)
curve for the same type of zeolite. Note continuous loss of water with
increasing temperature.

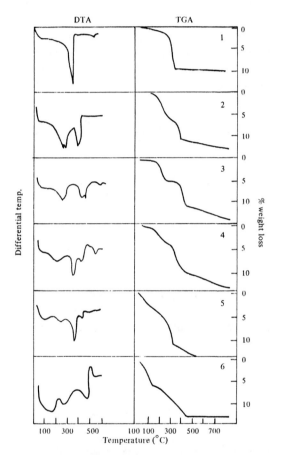

Figure 6. DTA and TGA curves for zeolites of the natrolite group. (1) Natrolite;
(2) mesolite; (3) scolecite; (4) thomsonite; (5) gonnardite; (6) edingtonite. (From Breck, 1974, Fig. 6.8.)

173

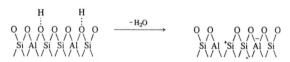

Modification of ammonium forms of zeolites involve complex structural and chemical reactions, including deammoniation, decationization, the formation of hydroxyl groups, framework dealumination, and framework stabilization. The latter two reactions are illustrated schematically below:

$$-O-\overset{\overset{O}{|}}{\underset{\underset{O}{|}}{Al}}-O-^{NH_4^+} \xrightarrow{3H_2O} -OH \quad \overset{H}{\underset{H}{\overset{O}{\underset{O}{}}}} HO- + Al(OH)_3 + NH_3$$

$$Al(OH)_3 + 2H^+ = Al(OH)^{+2} + 2H_2O$$

At higher temperatures,

$$Si(OH)_4 + -OH \quad \overset{H}{\underset{H}{\overset{O}{\underset{O}{}}}} HO- \longrightarrow -O-\overset{\overset{O}{|}}{\underset{\underset{O}{|}}{Si}}-O- + 4H_2O$$

The final, stabilized zeolite has a thermal and hydrothermal stability substantially higher than that of the starting material. It is believed that the increase in thermal stability is due to the formation of new Si-O-Si bonds as shown above.

REFERENCES

Ames, L.L. (1960) The cation sieve properties of clinoptilolite: Am. Mineral. 45, 689-700.

Ames, L.L. (1967) Zeolite removal of ammonium ions from agricultural waste-waters: Proc. 13th Pacific Northwest Industr. Waste Conf., Washington State Univ., 135-152.

Barrer, R.M. (1974) Isomorphous replacement by ion exchange: some equilibrium aspects: Bull. Soc. franc. Mineral. Cristallogr. 97, 89-100.

Barrer, R.M. (1978) Cation-exchange equilibria in zeolites and feldspathoids: In, Sand, L.B. and Mumpton, F.A., Eds., Natural Zeolites: Occurrence, Properties, Use, Pergamon Press, Elmsford, N.Y., 385-395.

Breck, D.W. (1974) Zeolite Molecular Sieves: Wiley-Interscience, New York, 771 pp.

Breck, D.W. and Smith, J.V. (1959) Molecular sieves: Scient. Am. 200, 85-94.

Eichorn, H. (1958) Ann. Phys. Chem. (Poggendorff) 105, 130.

Mercer, B.W., Ames, L.L., Touhill, C.J., Van Slyke, W.J., and Dean, R.B. (1970) Ammonia removal from secondary effluents by selective ion exchange: J. Water Poll. Cont. Fed. 42, R95-R107.

Chapter 9

UTILIZATION OF NATURAL ZEOLITES

F. A. Mumpton

INTRODUCTION

Historically, zeolitic tuffs have been used by man for more than 2000 years as lightweight dimension stone and in pozzolanic cements and concretes; however, it has been only within the last 25 years that the zeolite content of many of these materials has been recognized. The hundreds of discoveries of high-grade zeolite deposits in sedimentary rocks of volcanic origin since 1950 has led to the development of a host of applications which take advantage of the low mining costs of the near-surface deposits and of the attractive physical and chemical properties of the zeolite structures. Compared with the several hundred million pounds of synthetic molecular sieves sold each year throughout the world, the current use of natural zeolites in commercial applications other than those in the construction industry is relatively small; however, natural zeolites are in use today in small tonnages in the United States, Japan, and several other countries. Applications and potential applications range from the use as fillers in high-brightness papers, to the upgrading of low-BTU natural gas streams, to the removal of radioactive Cs^{137} from nuclear wastewaters, to dietary supplements for poultry and swine to produce larger animals at less cost.

The remainder of this chapter summarizes the current and potential uses that have been and are being developed for natural zeolites and has been taken largely from the author's article in "Natural Zeolites: Occurrence, Properties, Use," the conference volume of ZEOLITE '76. The author is extremely grateful to Pergamon Press, Inc. for their kind permission to use this material with but minor changes or revisions.

POLLUTION-CONTROL APPLICATIONS

In recent years zeolite minerals have found increasing application in the field of pollution abatement, and they are fast becoming standard components in the design and construction of such facilities. Both the ion-exchange and adsorption properties of zeolites can be utilized; however, most applications that have been developed are based on the ability of certain zeolites to exchange large cations selectively from aqueous solutions.

Radioactive-Waste Disposal

In 1959, L. L. Ames of the Hanford Laboratory of Battelle-Northwest, Richland, Washington, demonstrated the ion-exchange specificity of clinoptilolite for the removal of radioactive cesium and strontium from low-level waste streams of nuclear installations (Ames, 1959, 1960). The ions can be extracted with high efficiency from effluents and either stored indefinitely on the zeolite or removed by chemical means for subsequent purification and recovery. Using clinoptilolite crushed and screened to 20 x 50 mesh from the Hector, California, deposit, solutions containing the radioactive cations were passed through columns packed with the zeolite until breakthrough occurred. The "saturated" columns then were removed from the system and buried as solid waste. The process was actively investigated at Hanford and at several other nuclear-test stations in the United States in the 1960's (Knoll, 1963; Hawkins and Short, 1965), but supply problems and engineering reluctance to use the somewhat variable natural material that was available at that time hindered its complete acceptance by the industry. Millions of gallons of low-level Cs^{137} wastes, however, have been processed through zeolite ion exchangers at Hanford since that time, and a similar process was developed to recover this species from high-level effluents using a chabazite-rich ore from Bowie, Arizona (Nelson and Mercer, 1963; Mercer et al., 1970a). At the National Reactor Testing Station, Arco, Idaho, steel drums filled with granular clinoptilolite also were used as ion-exchange columns for Sr^{90} and Cs^{137} having half-lives of 25 and 33 years, respectively. Once the capacities of the drums were reached, they were removed, buried, and replaced with new drums containing fresh clinoptilolite (Wilding and Rhodes, 1963, 1965).

Similar processes have been investigated in several other countries, including Canada, Great Britain, France, Bulgaria, Hungary, Mexico, Japan, Germany, and the Soviet Union. Canadian workers devised a scheme for the removal and fixation of long-lived fission products using zeolite ion exchange (Mathers and Watson, 1962). More than 700 curies of Cs^{137} and Sr^{90} were extracted from 8200 liters of solution by 14 x 65 mesh clinoptilolite and placed in burial sites at Chalk River, Ontario. In Hungary, clinoptilolite from the Tokaj region has been used to encapsulate Sr^{90} for solid-waste disposal (Kakasy et al., 1973) and for the removal of both Sr^{90} and Cs^{137}

from low-level effluents (Adam et al., 1971). A mixture of vermiculite and clinoptilolite was found by Daiev et al. (1970) to take up more than 98% of the Cs^{137}, Tl^{204} As^{110}, Sr^{90}, and Ca^{45} in waste solutions, using zeolite ore from the large deposit at Kurdzali in southeastern Bulgaria. The use of clinoptilolite- or mordenite-rich tuff combined with a small amount of a non-ionic polymer coagulant was found to be particularly effective for Cs^{137} removal in Japan (Kato, 1974).

Clinoptilolite from the Georgian S.S.R. has been shown by Nikashina et al. (1974) to sorb strontium selectively from radioactive solutions. Vdovina et al. (1976) found clinoptilolite able to extract 94% of the radioactive Cs present in a waste stream containing 10g Cs/liter. According to a report of the International Atomic Energy Agency in Vienna (IAEA, 1972a), an altered volcanic tuff from the Eifel Region of Germany, marketed for more than 30 years as "filtrolit," is useful in the removal of Cs^{137} from waste streams. Tuffs rich in chabazite and phillipsite are known from this region, and it is likely that the tuff in question contains a large proportion of these minerals. The zeolite-rich Neopolitan Yellow Tuff from Naples, Italy, has also been used routinely to remove radionuclides from contaminated effluents at Casaccia, Italy (IAEA, 1972b). The abundance of chabazite and phillipsite in this tuff gives rise to its overall ion-exchange capacity for Cs of 2.1 meq/g and for Sr of 0.7 meq/g. The loaded tuff can be stored directly or incorporated into concrete for long-term burial.

Studies are currently in progress at the Los Alamos Laboratory of the University of California on the addition of powdered clinoptilolite along with flocculating agents to primary effluents to extract Cs^{137}. The solid material is later removed by anthracite filtration and placed in burial storage (Rohrer, 1976). The escalation of nuclear-power plant construction in this and other countries in the next two decades will result in the production of large quantities of radioactive wastes. Natural zeolites capable of extracting species such as Sr^{90}, Cs^{137}, Co^{60}, Ca^{45}, and Cr^{51} selectively in the presence of high concentrations of competing ions and which are capable of retaining their ion-exchange properties in high-flux environments may well play major roles in the safe development of nuclear power. Natural zeolites are not only considerably less expensive than organic ion-exchange resins, they are much more resistant to nuclear degradation. As silicates, they also react rapidly in cement or glass systems, entraining the radioactive species in the final concrete or vitreous products. An intriguing scheme being considered by several waste-disposal engineers consists of depositing drums of concentrated wastes or saturated ion-exchange columns in holes lined with several meters of packed clinoptilolite. The clinoptilolite would act as a filter-trap for small amounts of radioactive species that might leak from the drums or be leached from the glass or concrete in the years following burial.

Sewage-Effluent Treatment

As a spin-off of their work in radioactive waste disposal, Ames and Mercer showed that clinoptilolite is also highly selective for ammonium ions and suggested that it could be useful in the extraction of ammoniacal nitrogen from sewage and agricultural effluents (Ames, 1967; Mercer, 1969; Mercer et al., 1970b). Not only is NH_4^+ toxic to fish and other forms of aquatic life, it also contributes greatly to the rapid growth of algae and leads to eutrophication of lakes and streams. Increasingly stringent regulations on the amount of nitrogen that is permissible in municipal and industrial wastewater effluents have been promulgated in recent years by local and national environmental protection agencies. In general, a limit of about one part-per-million has been imposed, creating an urgent need to develop processes which will reduce the nitrogen content of streams to this level. Based upon test data from a mobile ion-exchange unit at Lake Tahoe in 1970 (Battelle Northwest, 1971) and subsequent engineering studies by Koon and Kaufman (1971), several large-scale sewage treatment plants were designed for communities in several parts of the United States. Plans call for clinoptilolite ion-exchange processes that would remove up to 99% of the contained ammonium ions from tertiary sewage effluents (Figure 1).

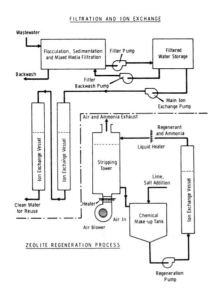

Figure 1. Flow chart of wastewater-treatment process using clinoptilolite ion exchange (Mercer et al., 1970b, Figure 7).

A 0.6 Mgd (million gallons per day) experimental unit went on stream at Rosemont, Minnesota, in the middle of 1975 and utilizes a total of 90 tons of 20 x 50 mesh clinoptilolite in six, 300 cu. ft. columns. In Virginia, 54 Mgd and 10 Mgd plants are in the construction stage for the Alexandria and Reston areas and will require 2000 and 400 tons of zeolite, respectively. The cost of the crushed and screened clinoptilolite is reported to be between $200 and $300 per ton, f.o.b. minesite. A 6 Mgd facility began operations in 1978 in the North Lake Tahoe Sewage District in California and uses several hundred tons of clinoptilolite (see Eyde, 1976). Two other units are contemplated for Waukegan, Illinois (Wilson, 1975) and Garabaldi, Oregon (Kapranos, 1976). The latter community overlooks Tillamook Bay, an important oyster production area in the Pacific Northwest, where special precautions are warranted to prevent contamination of the beds. A single physical-chemical unit for the removal of nutrients from combined sewer overflows in Syracuse, New York, has been designed by Murphy et al. (1978) and will employ a clinoptilolite ion-exchange process to extract NH_3-N.

If the performance of these first plants is satisfactory, the annual demand for clinoptilolite may exceed several hundred thousand tons in the United States alone within the next 20 years. During this period the country will also experience a sharp rise in the demand for water, giving added incentive for municipalities to conserve and reuse their existing water supplies. Studies in this area are in progress by CH_2M-Hill, Inc. (1975) for the Denver Board of Water Commissioners to develop a means to purify sewage effluent to potable standards. Preliminary plans have been proposed for a 1 Mgd prototype facility that will include a clinoptilolite ion-exchange step for removing the bulk of the ammoniacal nitrogen. Instead of releasing ammonia to the atmosphere during the regeneration steps, sulfuric acid will be added to produce ammonium sulfate to be sold locally as a fertilizer. An added benefit of clinoptilolite ion-exchange is that this zeolite appears to be able to extract trace amounts of heavy metals present in the wastewater which normally are detrimental to anticipated end uses (Chelischev et al., 1974; Hashimoto, 1974; Yoshida et al., 1976). The addition of zeolites to activated sludge also appears to aid oxidation and settling, according to Liles and Schwartz (1976).

In other countries the use of natural zeolites in wastewater treatment is only in the planning stages. The one exception is Japan where no less than fifty scientific articles and patents on the use of clinoptilolite and other zeolites for ammonium removal have been published since 1973. Some schemes call for powdered zeolite to be added to the effluent and then filtered or sedimented out (e.g., Sato and Fukagawa,

1976), while others employ ion-exchange columns filled with sized zeolite. According to Torii (1974), a 0.1 Mgd unit to remove nitrogen from the effluent of a soap and detergent factory has been in operation since 1971. Also, the 21,000 gallons of wastewater produced each day from a tourist observatory at Toba, Mie Prefecture, is now purified by means of zeolite ion exchange, using vessels packed with 2.5 tons of crushed clinoptilolite. Many more facilities of this type are planned in Japan for hotels, small factories, fish farms, and small communities lacking large-scale sewage-treatment systems.

Agricultural-Wastewater Treatment

In addition to the pollution from municipal and industrial wastewaters, streams and rivers in many parts of the United States are becoming more and more contaminated with nitrogen from irrigation runoff, seepage from animal feedlots, and waste streams from food processing plants. The use of natural zeolites to combat such pollution is a relatively untouched field, although some progress has been made, again mainly in Japan. Here farmers have sprinkled crushed zeolite on farmyard wastes for years (Minato, 1968), although the zeolitic nature of the adsorbent tuffs was only recently recognized. In the United States more than 700,000,000 tons of animal wastes are produced each day, and its disposal is becoming a major problem. The problem is especially acute in stockyards and dairy farms located close to large population centers where both air and water pollution must be rigidly controlled. Clinoptilolite appears to be doubly useful in the treatment of such materials, as it would not only remove most of the ammoniacal nitrogen from the liquid portion of the wastes and thereby decrease the ammonia-laden aerosols that travel many miles downwind of the feedlots and confining pens, but it would also retain much of this nitrogen in the solid form, thereby enhancing the fertilizer value of the manure. Such applications would most likely not require coarsely sized zeolite; powdered materials should be quite satisfactory and be readily available at prices of $50-75 per ton.

Stack-Gas Cleanup

Another area of pollution control involves the use of natrual zeolites in the removal of SO_2 and other pollutants from stack gases of oil- and coal-burning power plants About 25,000,000 tons of sulfur are dumped into the atmosphere each year by power stations in the United States, an amount that could double or triple in the next few years as the country turns to high-sulfur coals to allevaite its energy problems. Many federal and local air-pollution regulations now limit the amount of SO_2 in stack-gas

emissions to about 100 ppm; however, few processes now in use can meet these specifications. The relatively low concentrations of SO_2 in such emissions (about 3000 ppm) elude
efficient scrubbing operations, but zeolite adsorption might be an economical means of
concentrating this gas, as well as NO_x, CO_2, and various hydrocarbons for subsequent
removal (Blodgett, 1972). Certain natural mordenites and clinoptilolites are capable
of selectively adsorbing up to 200 mg of SO_2 per gram of zeolite under static conditions
and up to 40 mg/gram under dynamic conditions, even in the presence of copious amounts
of CO_2 (Ishikawa et al., 1971; Terui et al., 1974; Anurov et al., 1974; Smola et al.,
1975). They are especially suited to the low pH and high temperature conditions of
exhaust-gas systems (Roux et al., 1973). Linde Division of Union Carbide Corporation
has introduced a proprietary process for SO_2-cleanup of tailgases in sulfuric acid
plants based upon adsorption by an undisclosed molecular sieve (Miller, 1973).

Oil-Spill Cleanup

A novel use of zeolites in pollution control is as a sorbent in oil-spill cleanup.
Miki et al. (1974) pelletized a mixture of activated zeolite, expanded perlite, sodium
carbonate, tartaric acid, and a binder consisting of 20% methylsiloxane solution. The
product had a bulk density of 0.5 g/cc and an oil-sorption capacity of 0.97 g/gram. The
lightweight material was able to float on water for more than 200 hours and sorb oil from
the surface.

Oxygen Production

Air and water pollution generally involves the presence of objectionable compounds
and/or particles; however, they may also be caused by the absence of desirable ones,
such as oxygen. Oxygen deficiencies in lakes and streams result in the rapid extinction
of fish and plant life; in the atmosphere of a closed room its depletion is uncomfor-
table at best to human beings and hazardous at worst. Zeolite adsorption processes can
be utilized to produce inexpensive, oxygen-enriched streams of varying degrees of purity.
Based on earlier work of Barrer (1938), Domine and Haÿ (1968) showed that nitrogen gas
is selectively adsorbed from air by several zeolite materials yielding products containing up to 95% oxygen. A pressure-swing-adsorption process was developed in Japan
(Tamura, 1970) and has been in operation since 1968 at Toyohashi City, a few miles
north of Osaka, producing up to 500 m^3 of 90% O_2 per hour for use in secondary smelting
operations (Tamura, 1971). The plant consists of three towers, each filled with 13 tons
of acid-washed mordenite from the Itado Mine, Minase, Akita Prefecture, and operates at

room temperature on a nine-minute cycle of adsoprtion-desorption standby. It is reported to be competitive with air distillation in situations where enormous liquefaction facilities are not warranted. Portable and file-cabinet-size units have also been marketed in Japan to provide oxygen-enriched air for windowless restaurants where ventilation is poor. Larger units are in use to furnish oxygen to fish-breeding ponds and during the transportation of live fish (vide infra).

The potential markets for low-cost oxygen generators based on the selective adsorption properties of zeolite minerals are numerous and include river and pond aeration, pollution control in the paper and pulp industry by reoxygenation of downstream waters, and the generation of oxygen for secondary sewage treatment. The last application would find natural zeolites in direct conpetition with the more expensive synthetics currently used in several commercial systems, including Union Carbide's LINDOX process. This latter system is capable of producing up to 30 tons of 95% oxygen per day as the aeration gas in activated sludge systems (Union Carbide Corporation, 1975). The process has been successful for sewage-treatment plants operating at less than about 10 Mgd capacity.

Mordenite appears to be the best natural zeolite for oxygen generation, competing with the synthetic Ca-A zeolite, but certain clinoptilolites and chabazites also appear to be useful (Tamura, 1971; Torii et al., 1971, 1973; Tsitsishvili et al., 1972; Haralampiev et al., 1975). Hagiwara (1974) found that the selectivity of mordenite for nitrogen is enhanced by transformation of the zeolite to a mixed Na-H form.

By appropriate manipulation of adsorption and desorption cycles, these same zeolites can be used to prepare nitrogen streams with up to 99.95% purity. Although they are not strictly pollution-control applications, a number of uses for such inexpensive, high-purity nitrogen gas come to mind, including the topping-off of wine vats and beer barrels to prevent air from coming in contact with the liquid, the production of inert atmospheres in the storage and transportation of fruits and vegetables, the periodic flushing of home refrigerators to reduce spoilage and of silos to inhibit the rotting of ensilage and stored grains.

ENERGY-CONSERVATION APPLICATIONS

Coal Gasification

To cope with the world's growing energy requirements, whether they be from fossil fuels, nuclear, solar, or some other heretofore untapped source, modern technology will be forced to develop new processes or upgrade old ones to do a more efficient job. New or modified technology invariable calls for new or modified materials with properties

to match. Natural zeolites are a group of "new" materials that will undoubtedly be
thoroughly evaluated in the development of new energy sources or in the conservation of
old ones. In addition to their potential in the removal of SO_2 from stack-gas emissions
(vide supra), thereby allowing high-sulfur coals to be used in the production of elec-
tric power, the ability of certain zeolites to adsorb nitrogen selectively from air and
produce oxygen-rich products might find zeolites at the forefront of such fields as in-
place coal gasification. More than 25% of the coal reserves of the United States are in
deep-lying, unmineable seams and will require underground gasification if they are to be
utilized. Pumping air into the formations accelerates combustion but it also produces
large quantities of nitrogenous oxides and hydrocarbons which are dangerous and difficult
to handle. The use of oxygen-enriched air is preferred, but the cost of pure oxygen by
conventional liquefaction may be prohibitive. On-site zeolite adsorption units, however,
might be the answer in that that would be capable of producing any desired concentration
of oxygen at a lower price than liquefaction. In this manner combustion could be opti-
mized while minimizing equipment-corrosion problems that attend pure oxygen systems.

Natural-Gas Purification

Although the use of zeolites in coal gasification may be some years in the future,
natural zeolites have been employed successfully since 1968 to remove carbon dioxide
from contaminated and sour natural gas. Utilizing natural zeolites from its Bowie,
Arizona, deposit, NRG Corporation developed a pressure-swing-adsoprtion process (p.s.a.)
to extract up to 25% CO_2, H_2S, and H_2O from well gas in the Los Angeles area (see Wearout, 1971). With predicted shortages and rising prices of natural gas, attempts to
exploit heretofore unrecoverable sources become increasingly important. In 1975 this
company opened a methane-recovery and purification plant to treat gas produced by
decaying organic matter in the Palos Verde sanitary landfill in southern California.
Raw gas containing about 50% CH_4 and 40% CO_2 is routed through two pretreatment
vessels to remove moisture, hydrogen sulfide, mercaptans, etc. The dry gas is then
fed into three parallel adsorption columns packed with pellets of the chabazite/
erionite ore, and CO_2 is removed by p.s.a. techniques (Figure 2). Approximately one
million cubic feet of methane meeting pipeline specifications can be produced each day
and delivered to utility company pipelines (NRG, 1975).

That landfills generate methane gas has been known for years, but the CO_2 content
is usually too high for efficient use of the gas. Zeolite adsorption processes to
remove CO_2 and thereby raise the BTU content of the gas represent major breakthroughs
in this area. They should also open up many additional low-grade methane sources for

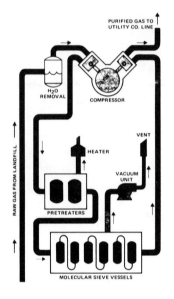

PURIFIED GAS TO
UTILITY CO. LINE

H₂O REMOVAL

COMPRESSOR

RAW GAS FROM LANDFILL

HEATER

VENT

VACUUM UNIT

PRETREATERS

MOLECULAR SIEVE VESSELS

Figure 2. Methane-purification scheme employed by Reserve Synthetic Fuels, Inc. at the Palos Verde landfill, Los Angeles, California. Natural chabazite from the Bowie, Arizona, deposit is used in a pressure-swing-adsorption process to adsorb CO_2 and other impurities. (From Mumpton, 1978, Fig. 16).

exploitation, such as the methane-rich gas produced in municipal sewage-treatment processes, in solid-waste incineration schemes, and during the digestion of feedlot manures Not only has the low BTU content of such gases limited their use, but H_2S and other acidic components present severe corrosion problems for valves and pumps. Zeolite adsorption rids the gas of these contaminants also.

Goeppner and Hasselmann (1974) estimated that as much as 2.5 ft³ of 70% methane gas is produced by the digestion of 100 gallons of municipal sewage. Commonly the impure product is used in-house as a source of heat for digestors or to operate pumps, but these uses tend to be inefficient and result in frequent equipment breakdown. If impurity CO_2 and H_2S could be removed from the gas, a 1000 BTU content could be prepared Similarly, the methane-rich gas formed by the digestion of animal wastes is also a potential source of energy, if the BTU content can be increased by CO_2 removal. Goeppner and Hasselmann (1974) also suggest that a billion cubic feet of 700 BTU gas could be produced by treating the 500 million pounds of manure produced each day by the 12,000,000 cattle now penned in feedlots throughout the United States. In this same area, Jewell (1974) indicates that if impurities could be removed, the methane produced by the digestion of the organic wastes on a typical New York State dairy farm of 60 cows could be equivalent to the farm's entire fossil-fuel requirements.

186

The production of methane gas from municipal solid wastes has been examined by Kispert et al. (1974), and it appears as if pure CH_4 could be produced at a cost of about $2.09 per thousand cubic feet by bacterial digestion, much like the anaerobic treatment of sewage sludge. A critical step in the proposed process is the scrubbing of the 50-50 methane-carbon dioxide product to remove CO_2. Zeolite adsorption would seem to be competitive with standard monoethanolamine adsorption, and costs might actually be less.

Solar Energy Use

Zeolites might also make contributions in the solar energy field. Detailed schemes for utilizing energy from the sun's rays are commonly stymied by the lack of efficient heat exchangers. During the last few years, considerable success has been achieved at the University of Texas and the Massachusetts Institute of Technology in using natural chabazite and clinoptilolite to adsorb and release heat from solar radiation for both air-conditioning and water heating. The dehydration of the zeolite by day and its rehydration at night results in the exchange of several hundred BTU's per pound of zeolite, sufficient to cool small buildings (Tchernev, 1978). Tchernev estimates that one ton of zeolite in solar panels spread over 200 square feet of roof surface will produce one ton of air conditioning. The extremely non-linear adsorption isotherms of crystalline zeolites, in contrast with other sorbent materials, make possible cooling efficiencies greater than 50%. If continued experimentation supports these initial results, solar energy uses may create a market for natural zeolites of several hundred thousand tons each year.

Petroleum Production

In addition to providing valuable information to exploration programs on the paleo-environments of suspected oil and gas fields, natural zeolites have been shown to be potentially useful in the purification and production side of the business. Generally, natural zeolites can not compete with certain synthetic molecular sieves in many adsorption and catalytic applications because of their inherently smaller pore sizes and adsorption capacities. Iron impurities also act as poison in many reactions. However, several natural zeolites have been useful in specific applications. Mordenite, chabazite, and clinoptilolite are generally capable of withstanding the rigors of continuous cycling in acid environments and have been used successfully to remove water and carbon dioxide from gaseous hydrocarbons (vide supra and Tsitsishvili et al., 1976). Union Carbide Corporation (1962) claims its chabazite-rich AW-500 product from Bowie, Arizona, to be effective in reformer-hydrogen drying, chlorine drying, chlorinated and fluorinated hydrocarbon purification, and in the removal of HCl from byproducts-hydrogen streams. Ohtani et al. (1972a, 1972b) showed that a modified mordenite could be used as a conversion catalyst in hydrocarbon disproportionation.

The selective forming process recently developed by Mobil R & D Corporation (Chen, 1971) makes use of catalysts derived from its erionite/clinoptilolite ore from Jersey Valley, Nevada. Neel et al. (1973) reported that erionite could be used as a carrier in the catalytic reforming of hydrocarbons. A Bulgarian clinoptilolite was shown by Nikolova and Ivanov (1975) to have potential in olefin hydrogenation, and Scherzer et al. (1975) recently patented a hydrocarbon cracking catalyst consisting of a mixture of H-Y zeolite and H-ferrierite, the latter from Lovelock, Nevada. Since 1970, several articles on the potential applications of clinoptilolite and mordenite from Tokaj, Hungary, as catalysts in cracking, isomerization, hydrogenation, alkylation, and polymerization have been published (Papp et al., 1971; Kallo et al., 1971; Detrekoy et al., 1974; Beyer, 1975). All make use of the hydrogen form of the zeolite.

AGRICULTURAL APPLICATIONS

Although farmers in Japan have used zeolitic tuffs for years to control the moisture content and malodor of animal manure and to raise the pH of acid volcanic soils, few data of a scientific nature have been published on these subjects. The increasing awareness of the abundance and availability of inexpensive natural zeolites has prompted considerable experimentation in the field of agriculture both in Japan and in the United States in the last ten years. Much of this work in the areas of animal science and aquaculture has been reviewed recently by Mumpton and Fishman (1977).

Fertilizers and Soil Amendments

Based upon their high ion-exchange capacity and water retentivity, natural zeolite have been used extensively in Japan as soil amendments, and small tonnages have been exported to Taiwan for this purpose (Minato, 1968; see also, Hsu et al., 1967). The pronounced ammonium selectivity of clinoptilolite has also been exploited in the preparati of chemical fertilizers which tend to improve the nitrogen-holding power of soils by increasing the bulk ion-exchange capacities and by promoting slower release of ammonium ions from the zeolite. In rice fields, efficiencies of less than 50% are not uncommon for surface-applied nitrogen; however, Minato (1968) reported a 63% improvement in the available nitrogen in a highly permeable paddy soil four weeks after 40 tons/acre zeolite was added with a standard fertilizer.

Turner (1975), on the other hand, noted little improvement in the overall nitrogen efficiency in field tests on a clay-rich rice soil in Texas, although laboratory tests showed increased ion-exchange capacity. He attributed these results to the fact that the Japanese soil contained much less clay, accounting for its inherent low ion-exchange capacity and fast draining properties. Thus, the addition of zeolite resulted in a marked improvement in nitrogen retentivity. These conclusions support those of Hsu et al. (1967) who found an increase in the effect of the zeolite when the clay content of the soil decreased.

Coupled with its valuable ion-exchange properties which allow a slow release of nutrients, such as Fe, Cu, Zn, Mn, and Co, to soils, clinoptilolite's ability to sorb excess moisture makes it an attractive additive to fertilizers to prevent caking and hardening during storage and to animal feedstuffs to inhibit the development of mold (Torii, 1974). In this area, Spiridonova et al. (1975) found that 0.5% clinoptilolite added to ammonium nitrate fertilizer decreased caking by 68%.

Pesticides, Fungicides, Herbicides

Similar to their synthetic counterparts, the high ion-exchange and adsorption capacities of many natural zeolites make them effective carriers of hericides, fungicides, and pesticides. Yoshinaga et al. (1973) found clinoptilolite to be an excellent carrier of benzyl phosphorothioate to control stem blasting in rice. Using natural zeolites as a base, Hayashizaki and Tsuneji (1973) developed a lime/nitrogen acaracide with good results. Mori et al. (1974) found that clinoptilolite is more than twice as effective as a carrier of the herbicide benthiocarb in eliminating weeds in paddy fields as other commercial products. Torii (1974) reported that more than 1000 tons of zeolite were used in Japan in 1975 as carriers in agriculture.

Heavy Metal Traps

Not only can the ion-exchange properties of certain zeolites be used to carry nutrient ions into soils, they can also be exploited to trap undesirable cations and lessen their introduction into the food chain. Fujii (1974) found that pulverized zeolites effectively reduced the transfer of fertilizer-added heavy metals, such as Cu, Cd, Pb, and Zn, from soils to plants. The selectivity of clinoptilolite for such heavy metals has been noted by several workers (e.g., Sato, 1975; Fujimori and Moriya, 1973; Chelishchev et al., 1974; Semmens and Seyfarth, 1978). In view of the many attempts being made by sanitary and agricultural engineers to add municipal and industrial sewage sludge to farm and forest soils, natural zeolites may play a major role in this area also. The nutrient content of such sludges is desirable, but the heavy metals present may accumulate to the point where they become toxic to plant life or to the animals or human beings that may eventually eat these plants. Cohen (1977) reported median values of 31 ppm Cd, 1230 ppm Cu, 830 ppm Pb, and 2780 ppm Zn for sludges produced in typical U. S. treatment plants. Zeolite additives to extract heavy metal cations from soils may be a key to the safe use of sludge as fertilizer and help extend the life of sludge-disposal sites and of land subjected to the spray-irrigation processes now being developed for the disposal of chlorinated sewage. Similarly, Nishita and Haug (1972) showed that the addition of clinoptilolite to soils contaminated with Sr^{90} resulted in a marked decrease in the uptake of strontium by plants, an observation having enormous import in potential treatment of radioactive fallout areas.

Animal Nutrition

Since 1965, experiments have been in progress in Japan on the use of clinoptilolite and mordenite as dietary supplements in animal husbandry. Up to 10% zeolite has been added to the normal rations of pigs, chickens, and ruminants, resulting in significant increases in feed-conversion values and in the general health of the animals. Onagi (1968) found that leghorn chickens on a diet containing 10% clinoptilolite gained as much weight as did birds on a normal diet, with no adverse effects on the vitality of the chickens or on their egg production. In addition, the droppings of test birds contained up to 27% less moisture, making the manure considerably easier to handle. Arscott (1975) found that broiler chickens fed a diet containing 5% clinoptilolite gained slightly less weight over a two-month period than birds on a normal diet, but that the average feed-conversion values (weight gain divided by nutritional feed intake) were significantly higher. Of equal or greater import was the fact that none of the 48 test birds died during the experiment, whereas three birds on the control diet and two on a control + antibiotic diet succumbed.

Kondo and Wagai (1968) observed similar results with swine. They found that pigs on diets containing 5% clinoptilolite gained up to 25% more weight than those on normal diets. No toxic or other adverse effects were noted. On the contrary, the presence of zeolite in the rations appeared to contribute measurably to the well being of the animals. According to Torii (1974), the death rate and incidence of sickness among swine fed a diet containing 6% clinoptilolite was markedly lower than for control animals in experiments conducted on a large swine-raising facility in Japan. Scours and other intestinal diseases were greatly reduced. Apparently, the vitalizing effect of a zeolite diet can also be transferred from mother to offspring. Torii (1974) reported that the growth rate of newborn pigs whose mothers were fed 600 g of clinoptilolite each day of the 35-day weaning period was 65-85% greater than those of control animals. In addition, the young pigs in the test group suffered almost no attacks of diarrhea, as opposed to those in control groups who were severely afflicted with scours, inhibiting their normal growth.

Similar studies were conducted at Oregon State University (England, 1975) on young pigs using rations containing 5% clinoptilolite from the Hector, California, deposit. Although lesser increases in growth rate were observed, the incidence of scours was significantly reduced for animals receiving the zeolite diet. Currently, heavy doses of prophylactic antibiotics are used to control intestinal diseases which, unchecked, can result in high mortality rates in young swine after they are weaned. If federal regulations prohibit such usage, other means must be found, and natural zeolites might be the answer. The exact functions of the zeolite in both dietary and antibiotic phenomena are not well understood and await serious physiological and biochemical explanation. The ammonium selectivity of clinoptilolite suggests that it might act as a nitrogen reservoir in the digestive system of the animal, allowing a slower release and more efficient use of ammonium ions produced by the breakdown of ingested rations in the development of animal protein. Zeolite particles might also stimulate the linings of the stomach and intestinal tract causing the animal to produce more antibodies, and thus inhibiting diseases such as scours.

In an attempt to reduce the toxic effects of high NH_4^+ concentrations in ruminal fluids when non-protein nitrogen compounds (NPN), such as urea and diuret, are added to the diets of cattle, sheep, and goats, White and Ohlrogge (1974) introduced both natural and synthetic zeolites into the rumen of test animals. Ammonium ions formed by the enzyme decomposition of NPN were immediately ion exchanged into the zeolite structure and held there for several hours until released by the regenerative action of sodium ions entering the rumen in saliva. Thus, the gradual release of ammonium ions allowed rumen micro-organisms to synthesize cellular protein continuously for easy assimulation into the animals' digestive systems. The zeolite was introduced as pellets or beads in a foraminous bolus or as a finely divided powder mixed with the normal rations. Both natural chabazite and clinoptilolite were found to be effective, although synthetic zeolite F worked best of all. Considering the possible price of about $1.00 per pound for zeolite F, if and when it becomes commercially available, it would seem that certain NH_4^+-selective natural zeolites, such as clinoptilolite, would be especially useful in this application. Prices of a few pennies a pound for the natural zeolites suggest that they could be fed to the animal as part of its normal rations.

In this same area Kondo et al. (1969) found that zeolites added to the feed improved the growth rate of calves by stimulating their appetite and improving digestion. Watanabe et al. (1971) raised six bullocks for 329 days on a diet containing 2% clinoptilolite. Little difference in weight gain or feed conversion was noted between these control animals, but the test steers showed slightly larged body dimensions and dressed out to give higher quality meat. The differences were reflected in a 20% larger profit compared to animals on a normal diet. Diarrhea and other intestinal diseases were also noticeable in the test animals. It is unfortunate that a much higher level of zeolite was not used in these experiment; earlier experiments in the United States added as much as 40% clay to the rations of domestic animals without adverse effects (Ousterhout, 1970).

Excrement Treatment

The rising demand for animal protein in worldwide diets is attended by several major problems, not the least of which is what to do with the mountains of animal wastes produced each day on farms and in feedlots of every country. Livestock production in the United States alone creates more than one billion tons of solid waste and nearly 400,000,-000 tons of liquid waste each year (Laporte, 1975, p. 365). Accumulations of such materials pose serious health problems to man and animal alike and are significant sources of pollution for rivers and streams in the area. Large amounts of undigested protein remaining in the excrement represents a valuable resource that for the most part is now wasted, due to our growing dependence on chemical fertilizers. Natural zeolites have potential application in several areas of manure treatment, including (1) reducing the malodor of the excrement, (2) controlling the moisture content for ease of handling, and (3) purifying methane gas produced by the anaerobic digestion of such materials (vide supra).

The semi-fluid droppings in large poultry houses commonly emit a stench that is discomforting to farm workers and to the chickens themselves. Lower egg production and smaller animals are common results, as many birds suffer from respiratory diseases caused by the noxious fumes of ammonia and hydrogen sulfide. Torii (1974) reported that about 100 tons of clinoptilolite are used each year in the chicken houses of Japan. The zeolite is either mixed with the droppings or packed in boxes which are suspended from the ceilings. In either case, ammonia is removed, and egg production is increased. Similar results were obtained by Onagi (1966) and Kondo and Wagai (1968) when zeolites were added to the rations of chickens and pigs, respectively. About 25 tons of clinoptilolite are spread on the floors of a Sapporo pig farm each month to adsorb urine and other liquid wastes. The buildings are said to be dry, clean, and considerably less odoriferous. An innovative use of natural zeolites in the treatment of animal excrement was developed by Komakine (1974) and involves mixing chicken droppings with zeolite and ferrous sulfate. The ferrous sulfate inhibits zymosis and decomposition of the droppings, while the zeolite stabilizes the hygroscopic nature of this compound and captures ammonium ions produced by the manure. The mixture is dried and can be used as an odorless fertilizer, or it can be used as an odorless feedstuff, rich in protein, for fowl, fish, and domestic animals. It has been used successfully in Japan to substitute for as much as 20% of the normal rations of swine, poultry, and edible carp.

Aquacultural Uses

Aquaculture may not be the number-one food industry in the United States nor in very many other nations for that matter, but as the world's protein requirements continue to rise, more and more fish products may find their way to the dinner table or to the feeding trough in future years. Although progressive farmers in several countries are beginning to learn that catfish or trout can bring in more dollars per acre than wheat or rice, fish raising is a tricky business, and the chemical and biological environment of aquacultural systems must be maintained within close limits at all times. Natural zeolites have been found to be valuable materials in controlling these environments, and as more fish hatcheries turn to self-contained, water-reuse systems, a sizeable market for natural zeolites may develop.

Using processes similar to those employed in sewage-treatment plants for the extraction of ammoniacal nitrogen, Konikoff (1974) and Johnson and Sieburth (1974) showed clinoptilolite to be effective in the removal of ammonium ions from recirculating fish-culture systems. Ammonium is one of the most significant toxic metabolites in aquaculture and is extremely harmful to fish in concentrations exceeding a few parts-per-million. In oxygen-poor environments such concentrations can lead to damage of gill tissue, various gill diseases, and a reduction in the growth rate (Larmoyeaux and Piper, 1973). Unpublished results of tests conducted in 1973 at a working hatchery near Newport, Oregon, indicated that 97-99% of the ammonium produced in a recirculating system was removed by clinoptilolite ion-exchange columns from waters containing 0.34-1.43 mg NH_3-N per liter (Kapranos, 1976). Peters and Bose (1975) substantiated these results and found that trout remained perfectly healthy during a four-week trial when zeolite ion exchange was used to remove ammonium from recirculating tank water. They concluded that selective ion exchange using clinoptilolite may be a viable alternative to biological oxidation processes which are highly susceptible to minor changes in the temperature and chemistry of such systems. At the present time Becker Industries of Newport, Oregon, in conjunction with the U. S. Army Corps of Engineers, is developing a complete

193

purification unit for hatchery-water reuse. The system will incorporate a zeolite ion-exchange unit for ammoniacal nitrogen removal and is designed to handle concentrations of 20-30 ppm NH_3-N at flow rates of 10-15 Mgd, typical of the more than 200 fish hatcheries now operating in the Pacific Northwest (Kapranos, 1976).

Jungle Laboratories of Comfort, Texas, is in the process of developing ammonia-removal systems for hatchery waters using clinoptilolite ion exchange and for fish haulage operations where brain damage due to excess ammonium ions commonly results in sterility, stunted growth, and high mortality rates (Nichols, 1976). Throw-away cartridges and filters containing granular clinoptilolite are also being designed to regulate the ammoniacal nitrogen content of home aquaria and tanks used to transport tropical fish from the time they are caught until they reach the hobbyist. The U. S. Fish and Wildlife Service has investigated similar processes for ammonia removal from recirculating waters in tank trucks used to transport channel catfish from Texas to the Colorado River in Arizona (McCraren, 1976). Normally about 3500 pounds of 8-10" fish are carried in 12,000 gallons of water, or about 2.2 lbs/ft^3. If the ammonium ions produced by decomposing fish excrement and waste food can be extracted, the number of fish hauled in trucks of this size can be doubled.

Slone et al. (1975) were able to achieve biomass concentrations of more than 7 lbs/ft^3 for catfish (Ictaluras punctatus) after seven months in a vertical raceway system using recycled water and a zeolite ion-exchange process to remove ammoniacal nitrogen following biofiltration. NH_4-N concentrations of 5 mg/l were maintained easily with clinoptilolite from the Hector, California, deposit.

As discussed above, zeolite adsorption units using natural mordenite are currently marketed in Japan and have been used to provide oxygen for aeration in fish culture and in the transportation of live fish (Minato, 1974). Small generators capable of producing 15 liters of 50% O_2 per hour are manufactured by the Koyo Development Company, Ltd. Carp and goldfish raised in oxygen-aerated water are said to be livelier and to have greater appetites (Koyo Kaihatsu, 1974). Little work has been carried out on the use of natural zeolites in fish food, but the protein-rich mixture of chicken droppings, ferrous sulfate, and zeolite discussed above (Komakine, 1974) has been fed to carp in Japan with no adverse effects. Much as the excrement of pigs, chickens, and cattle is rendered less odoriferous by the addition of clinoptilolite to the rations of the animals, it would seem that natural zeolites incorporated into normal fish food would reduce the ammonia buildup in closed tanks or in large holding facilities where water is recirculated. Such possibilities bear investigating.

In the area of fish processing, Araki and Honda (1974) found that zeolitic tuff is able to deodorize protein processing odors. Gas containing large proportions of ammonia, volatile amines, mercaptans, and acetic acid were purified when passed through a column of pelletized zeolite at 250-400°C.

MINING AND METALLURGY APPLICATIONS

Exploration Aids

Recent studies in Japan indicate that zeolite assemblages in altered tuffs can not only delineate the conditions of formation of certain ore deposits but also serve as exploration tools, especially in areas of thick overburden. Yoshida and Utada (1969) and Utada et al. (1974) noted that analcime-rich aureoles surrounding Kuroko-type mineralization in Neogene sediments of the Green Tuff region are thickest in the vicinity of major ore deposits. Supported by the experimental results of Aoki (1974) on the conversion of clinoptilolite to analcime in sodium-carbonate solutions, they suggest that hydrothermal alteration was superimposed on zeolitic alteration that took place earlier during burial diagenesis of the marine-tuff host rocks. The thickness of analcimic alteration in such rocks, therefore, may be a clue to the location of deep-seated ore bodies.

Also in Japan, Katayama et al. (1974) attributed the concentration of uranium to the presence of a heulandite-clinoptilolite zeolite in tuffaceous sandstones of Miocene age near Tono, Gigu Prefecture. Oxidized uranium in ground water is presumed to have been adsorbed on the zeolite, which in some zones contains as much as 0.9% U. Although considerable experimentation is still required, these studies suggest that natural zeolites may be used to extract and concentrate uranium and other ions from low-level processing solutions, such as those encountered during the in-place or heap-leaching of uranium and copper ores and tailings.

Metallurgical Uses

Similarly, the ion-exchange properties of many natural zeolites lend themselves to the concentration of heavy metals from wastewater effluents of mining and metallurgical operations. Torii (1974) reported that sodium-exchanged clinoptilolite and mordenite removed almost all of the Cd^{2+} from 10-ppm solutions of that ion. Mondale et al. (1977) also pointed out the effectiveness of certain zeolites for extracting heavy metal ions from aqueous media. They found that chabazite is able to remove more than 90% of the Pb^{2+} and Cu^{2+} from 10^{-3} molar solutions. Chelishchev et al. (1974) successfully extracted large, heavy-metal cations from complex solutions using clinoptilolite from the

Georgian S.S.S.R. They reported a selectivity of Pb > Ag > Cd, Zn, Cu > Na. Fujimori and Moriya (1973) noted that clinoptilolite removed heavy metal ions from industrial wastewater in the sequence Pb >> Cd > Cu >> Zn. Although the overall specificities of individual zeolites for heavy metals have not yet been established, the preliminary data suggest that natural zeolites can contribute heavily towards eliminating a large part of the pollution from industrial wastewaters and also provide a means of recovering heretofore lost values from many hydrometallurgical effluents.

In pyrometallurgy, Fusamura et al. (1976) found that a combination of $CaCO_3$ and natural zeolite suppressed up to 90% of the lead fumes from molten Cu-Pb alloys, when the mineral mixture was floated on top of the liquid metals.

MISCELLANEOUS APPLICATIONS

Paper Products

Approximately 3000 tons of high-brightness zeolites from the Itaya Mine, Yamagata Prefecture, Japan, is sold each month by the Zeeklite Chemical Company under the trade names SGW and HiZ as filler in the paper industry (Minato, 1975). Clinoptilolite ore is finely ground and classified by wet or dry cycloning into a -10 micrometer-size product having an abrasion index of less than 3% and a brightness of about 80 (Takasaka, 1975). According to Kobayashi (1970), kraft papers filled with clinoptilolite are bulkier, more opaque, easier to cut, and less susceptible to ink blotting than those filled with clay. Hayakawa and Kobayashi (1973) patented a lightweight paper made from a mixture of 17% pulp, 25% chemically ground pulp, 28% bleached kraft, 1% sizing, 1% aluminum sulfate, and 28% zeolite powder. The product has a density of 0.68 g/cc compared with 0.73 g/cc for conventional paper. A few hundred tons of clinoptilolite from the Tokaj district of eastern Hungary (Nemecz and Varju, 1962) are also mined each month for paper-filling applications, although Kobor and Hegedus (1968) found Hungarian natural zeolites to be unsuited for the manufacture of wood-free papers.

Kato (1976) combined clinoptilolite and mordenite with organic dyes, such as C.I. Basic Red 34, to give heat-, light-, and acid-resistant coloring composites useful in coating copying papers and in coloring plastics. Breck (1975) found that the addition of from 5-30% natural chabazite or synthetic A or X zeolites to conductive paper used in electrostatic reproduction resulted in a product having resistivities of 10^7 to 10^8 ohm-centimeters. The strong hydration properties of the zeolites allowed these values to be maintained over a range of relative humidities of from 5 to 90%.

Construction Uses

Much as perlite and other volcanic glasses can be expanded into low-density pellets for use as lightweight aggregate in cements and concretes, natural zeolites also can be frothed into stable products by calcining at elevated temperatures. Stojanović (1972) reported densities as low as 0.8 g/cc and porosities of up to 65% for expanded clinoptilolite from several Serbian deposits after firing at 1200-1400°C. Similar materials were prepared by Ishimaru and Ozata (1975) from Japanese zeolites at 1250°C. Kasai et al. (1973) prepared foaming agents by calcining clinoptilolite at about 550°C and cooling in air. The product was mixed with equal parts water and dolomite plaster, molded, and hardened for two hours in an autoclave to obtain a product having a bulk density of 0.75 g/cc and a compressive strength of about 47 kg/cm². In a search for perlite substitutes, Bush (1974) found that high-grade clinoptilolite from the Barstow Formation, California, expanded 4 to 6 times when heated for 5 minutes at from 1150-1250°C. Synthetic CaA zeolite loaded with 1-butene or butadiene was used by Ulisch (1975) to prepare a porous foam having a density of 0.95 g/cc; natural erionite or chabazite should work equally as well. The temperatures required for expanding zeolitic tuffs is significantly higher than those needed for perlite or other expandable materials (1200°C vs 760°C); however, the foamed zeolite products are considerably stronger and more resistant to abrasion.

Also in the construction area, Torii (1974) reported that about 100 tons of clinoptilolite are consumed each month as filler in wheaten paste used to adhere plywood prior to hot pressing. He also listed 200 tons per month of mordenite as being used to fabricate lightweight bricks (0.7-1.0 g/cc) of high physical strength and high acid- and alkali-resistance. By firing a mixture of 100 parts glass, 70 parts clinoptilolite, 3 parts carbon, and 3 parts H_3PO_4 at 800°C, Tamura (1974) was able to produce a high-strength porous glass having a density of only 0.22 g/cc.

Medical Applications

In the medical field, Kato et al. (1969, 1970) found that clinoptilolite is useful as a polishing agent in fluoride-containing toothpaste. It is no more abrasive than the commonly used $CaHPO_4$ and allows more of the fluoride ion to remain in the anionic form. Andersson et al. (1975) were able to separate ammoniacal nitrogen from hemodialysis liquids in recycle-dialysis systems using a natural phillipsite. The zeolite was found to be superior in selectivity and ion-exchange capacity to other exchangers, including zirconium phosphate, Dowex 8x50, clinoptilolite, and Union Carbide Corporation's AW-500 chabazite/erionite product.

197

ACKNOWLEDGMENTS

Literature reviews of rapidly expanding subjects, such as the geological signifi-cance and the industrial utilization of natural zeolites, can never be up to date; however, with the generous help of many individuals in academe, industry, and govern-ment, this author has tried to make this review as current as possible. The author's gratitude is extended to L. L. Ames, G. H. Arscott, F. F. Aplan, D. C. England, W. Kapranos, M. Koizumi, R. B. Laudon, R. T. Mandeville, H. Minato, E. Nemecz, L. B. Sand, R. A. Sheppard, D. Tchernev, and D. E. W. Vaughan for numerous discussions and for pro-viding unpublished information. Special thanks are due Mr. Kazuo Torii of the National Industrial Research Institute, Tohoku, Sendai, Japan, for his many courtesies in pro-viding data on the use of natural zeolites in Japan and for his translations of several articles.

REFERENCES

Adam, L., Kakasy, Gy., and Pallos, I. (1971) Removal of radioactive pollutants from water using mineral substances of natural domestic origin: Hung. Mining Res. Inst. 14, 209-212.

Ames, L.L. (1959) Zeolitic extraction of caesium from aqueous solutions: Unclass. Rep. HY-62607, U.S. Atomic Energy Comm., 23 pp.

Ames, L.L. (1960) The cation sieve properties of clinoptilolite: Am. Mineral. 45, 689-700.

Ames, L.L. (1967) Zeolitic removal of ammonium ions from agricultural wastewaters: Proc. 13th Pacific Northwest Indust. Waste Conf., Washington State Univ., 135-152.

Andersson, S., Grenthe, I., Jonsson, E., and Naucler, L. (1975) Separation of poisons from the dialysis liquid of a recycle dialysis system: Ger. Offen. 2,512,212, Sept. 25, 1975, 30 pp.

Anurov, S.A., Keltsev, N.V., Smola, V I., and Torocheshnikov, N.S. (1974) Adsorption of SO_2 by natural zeolites: Zh. Fiz. Khim. 48, 2124-2125.

Aoki, M. (1974) Synthesis of analcime from clinoptilolite tuff in $NaCO_3$ solution. Experimental consideration related to the formation of analcime zone surrounding Kuroko deposits: Ganseki Kobutsu Kosho Gakkaishi 69, 171-180.

Araki, K. and Honda, S. (1974) Deodorizing process: Japan. Kokai 74,034,898, Sept. 18, 1974, 2 pp.

Arscott, G.H. (1975) Dept. of Poultry Science, Oregon State University, Corvallis (personal communication).

Barrer, R.M. (1938) The sorption of polar and non-polar gases on zeolites: Proc. Roy. Soc. Lond. 167A, 392-419.

Battelle-Northwest (1971) Wastewater ammonia removal by ion exchange: Rep. Project No. 17010 ECZ for U.S. Environ. Protect. Agency, 62 pp.

Beyer, H. (1975) Cracking reactions. II. Kinetics of the catalytic cracking of propane, butane, and isobutane on H-clinoptilolite: Acta Chim. Acad. Sci., Hung. 84, 25-43.

Blodgett, G.A. (1972) SO_2 adsorption on ion-exchanged mordenites: M.S. thesis, Worcester Polytechnic Institute, 96 pp.

Breck, D.W. (1975) Electrostatic printing process: U.S. Patent 3,884,687, May 20, 1975, 5 pp.

Bush, A.L. (1974) National self-sufficiency in lightweight aggregate resources: Minutes - 25th Annu. Meet. Perlite Inst., Colorado Springs, Colorado, April 18-23, 1974.

Chelishchev, N.F., Martynova, N.S., Fakina, L.K., and Berenshtein, B.G. (1974) Ion exchange of heavy metals on clinoptilolite: Dok. Akad. Nauk S.S.S.R. 217, 1140-1141.

Chen, N.Y. (1971) Shaped selective transition metal zeolite hydrocracking catalysts: U.S. Patent 3,630,966, Dec. 28, 1971.

CH_2M-Hill (1975) Conceptual design report--potable water reuse plant, successive use program: for Board of Water Commissioners, Denver, Colorado, Vol. I, Vol. II.

Cohen, J.M. (1977) Trace metal removal by wastewater treatment: Tech. Transfer, U.S. Environ. Protect. Agency, Jan. 1977, 2-7.

Daiev, Ch., Delchev, G., Zhelyazkov, V., Gradev, G., and Simov, S. (1970) Immobilization of radioactive wastes attached to natural sorbents in bitumen molds: Int. Atomic Energy Agency, Vienna, Symp. Management of Low- and Intermediate-level Radioactive Wastes, 739-746.

Detrekoy, E.J., Jacobs, P.A., Kallo, D., and Uytterhoeven, J.B. (1974) Catalytic activity of hydroxyl groups in clinoptilolite: J. Catalysis 32, 442-451.

Dominé, D. and Haÿ, L. (1968) Process for separating mixtures of gases by isothermal adsorption: possibilities and application: In, Molecular Sieves, Soc. Chem. Indus., 204-216.

England, D.C. (1975) Effect of zeolite on incidence and severity of scouring and level of performance of pigs during suckling and early postweaning: Rep. 17th Swine Day, Spec. Rep. 447, Ag. Ex. Stat., Oregon State Univ., 30-33.

Eyde, T.H. (1976) Zeolites: Min. Eng., March, 51-53.

Fugii, S. (1974) Heavy metal adsorption by pulverized zeolites: Japan. Kokai 74,079,849, Aug. 1, 2 pp.

Fujimori, K. and Moriya, Y. (1973) Removal and treatment of heavy metals in industrial wastewater. I. Neutralizing method and solidification by zeolite: Asahi Garasukogyo Gijutsu Shoreikai Kenkyu, Hokoku 23, 243-246.

Fusamura, N., Nagoya, T., Oya, S., Takada, T., Furuta, S., and Fujii, T. (1976) Effects of covering conditions on the suppression of lead fume from molten Cu-Pb alloys: Imono 48, 512-519.

Goeppner, J. and Hasselmann, D.E. (1974) Digestion by-product may give answer to energy problem: Water and Wastes Eng. 11, 30-35.

Hagiwara, Z. and Yamamoto, T. (1974) Selective enrichment of oxygen and nitrogen: Japan. Kokai 74,054,289, May 27, 9 pp.

Haralampiev, G.A. et al. (1975) Possibilities of oxygen enrichment of air for metallurgy using Bulgarian clinoptilolite: Annu. Rep. HIC-HCTI-Sofia, 60-75.

Hashimoto, A. (1974) Adsorption of heavy metals and application to watersupply systems.
II. Removal of Ca, Cd, and N as ammonia: Mizu shori Gijutau 15, 1085-1091.

Hashimoto, S. and Miki, K. (1975) Water-purifying materials: Japan. Kokai 75,098,044,
May 1, 5 pp.

Hawkins, D.B. and Short, H.L. (1965) Equations for the sorption of cesium and strontium
on soil and clinoptilolite: U.S. Atomic Energy Comm. IDO-12046, 33 pp.

Hayakawa, J. and Kobayashi, T. (1973) Fillers for paper: Japan. Kokai 73,099,402,
Dec. 15, 3 pp.

Hsu, S.C., Wang, S.T., and Lin, T.H. (1967) Effects of soil conditioners on Taiwan
soils. I. Effects of zeolite on physico-chemical properties of soils: J. Taiwan
Ag. Res. 16, 50-57.

IAEA (1972a) Use of local minerals in the treatment of radioactive wastes: Int. Atomic
Energy Agency, Vienna, Tech. Rep. 136, 97-98.

IAEA (1972b) Use of local minerals in the treatment of radioactive wastes: Int. Atomic
Energy Agency, Vienna, Tech. Rep. 136, 68, 98.

Ishikawa, H., Tanaka, H., Uchiyama, K., Kwon, S., Asayama, T., and Morimoto, T. (1972)
Adsorption and desorption of SO_2 gas on natural zeolites: Waseda Diag. R.K. Kark.
Hok. 51, 46-53.

Ishimaru, H. and Ozata, K. (1975) Ceramic foam: Japan. Kokai 75,028,510, March 24,
5 pp.

Jewell, W.J. (1974) Methane. . .the energy-sufficient farm: State Univ. New York -
The News 3, no. 2, p. 4.

Johnson, P.W. and Sieburth, J.M. (1974) Ammonia removal by selective ion exchange, a
backup system for microbiological filters in closed-system aquaculture: Aquacul.
4, 61-68.

Kakasy, I., Pallos, I., and Adam, L. (1973) Étude des roches clinoptilolitiques de la
Hongrie au point de vue de la fixation des déchets radioactifs liquides: Publ.
Hung. Min. Res. Inst. No. 16, 169-175.

Kallo, D. et al. (1971) Catalytic isomerization of butene: Kem. Kozlem. 36, 239-245.

Kapranos, W. (1976) Becker Industries, Portland, Oregon (personal communication).

Kasai, J., Urano, T., Kuji, T., Takeda, Y., and Machinaga, O. (1973) Inorganic foaming
agents by calcination of natural silica-high zeolites: Japan. Kokai 73,066,123,
Sept. 11, 5 pp.

Katayama, N., Kubo, K., and Hiorno, S. (1974) Genesis of uranium deposits of the Tono
Mine, Japan: Proc. Symp. Formation of Uranium Ore Deposits, Athens, Greece,
May 6-10, 1974, 437-452.

Kato, Chuzo (1976) Coloring composites: U.S. Patent 3,950,180, April 13, 5 pp.

Kato, K., Nagata, M., Sakai, Y., Shiba, M., and Onishi, M. (1969) Zeolite as polishing
agent for dentrifice I. (Chemical) properties of zeolite: Iwate Daigaku Nagakubu
Hokoku 3, 22-35.

Kato, K., Nagata, N., Sakai, Y., Shiba, M., Kojima, M., and Onishi, M. (1970) Zeolites
as polishing agent for detrifice. II. Abrasive action and active fluoride ion in
zeolite-containing fluoride: Iyokizikai Kenkyusho Hokoku 4, 115-128.

Kato, K. (1974) Removal of Cs^{137} from aqueous solutions by zeolite: Hoken Butsuri 9, 11-16.

Kispert, R.G., Anderson, L.C., Walker, D.H., Sadek, S.E., and Wise, D.L. (1974) Fuel gas production from solid waste: NSF/RANN/SE/C Rep. 827/PR/74/2, 176 pp.

Knoll, K.C. (1963) Removal of cesium from aqueous solutions by adsorption: U.S. Patent 3,097,920.

Kobayashi, Y. (1970) Natural zeolite-fillers for paper: Japan. Kokai 70,041,044, 2 pp.

Kobor, L. and Hegedus, I. (1968) Hungarian zeolite in the paper industry: Papiripar 12, 44-50.

Komakine, C. (1974) Feedstuff for fowl, fish, and domestic animals: U.S. Patent 3,836,676, Sept. 17, 3 pp.

Kondo, K., Fujishiro, S., Suzuki, F., Taga, T., Morinaga, H., Wagai, B., and Kondo, T. (1969) Effect of zeolites on calf growth: Chikusan No Kenikyu 23, 987-988.

Kondo, N. and Wagai, B. (1968) Experimental use of clinoptilolite-tuff as dietary supplements for pigs: Yotonkai, May, 1-4.

Konikoff, M.A. (1973) Comparison of clinoptilolite and biofilters for nitrogen removal in recirculating fish culture systems: Ph.D. dissertation, Southern Illinois Univ., Univ. Microfilms No. 74-6222, 98 pp.

Koon, J.H. and Kaufman, W.J. (1971) Optimization of ammonia removal by ion exchange using clinoptilolite: Water Pollution Control Research Series 17080, DAR 09/71, 189 pp.

Koyo Kaihatsu Co., Ltd. (1974) Adsorption Oxygen Generator: Company Brochure, 4 pp.

Larmoyeaux, J.D. and Piper, R.G. (1973) Effects of water reuse on rainbow trout in hatcheries: Prog. Fish Cult. 35, 2-8.

Laporte, L.F. (1975) Encounter with the Earth: Canfield Press, San Francisco, 538 pp.

Liles, A.W. and Schwartz, R.D. (1976) Method of treating wastewater: U.S. Patent 3,968,036, July 6, 10 pp.

Mathers, W.G. and Watson, L.C. (1962) A waste disposal experiment using mineral exchange on clinoptilolite: Atomic Energy of Canada, Ltd., AECL-1521.

McCraren, J. (1976) U.S. Fish and Wildlife Services, San Marcos, Texas (personal communication).

Mercer, B.W. (1969) Clinoptilolite in water-pollution control: Ore Bin 31, 209-213.

Mercer, B.W., Ames, L.L., and Smith, P.W. (1970a) Cesium purification by zeolite ion exchange: Nucl. Appl. & Tech. 8, 62-69.

Mercer, B.W., Ames, L.L., Touhill, C.J., Van Slyke, W.J., and Dean, R.B. (1970b) Ammonia removal from secondary effluents by selective ion exchange: J. Water Poll. Cont. Fed. 42, R95-R107.

Miki, K., Oyama, R., and Kitagawa, H. (1974) Oilspill adsorbent containing zeolite and perlite: Japan. Kokai 74,007,184, April 12, 1972, 6 pp.

Miller, W.C. (1973) Adsorption cuts SO_2, NO_x, Hg: Chem. Eng., Aug. 6, 62-63.

Minato, H. (1968) Characteristics and uses of natural zeolites: Koatsugasu 5, 536-547.

Minato, H. (1974) Properties of sedimentary zeolites for industrial application: Seminar on the Occurrence, Origin, and Utilization of Sedimentary Zeolites in the

Circum-Pacific Region U.S.-Japan Cooperative Science Program, Menlo Park, California, July (unpublished abstract).

Minato, H. (1975) Zeolite, its natural resources and utilization: Ceramics (Japan) 10, 914-957.

Mondale, K.D., Mumpton, F.A., and Aplan, F.F. (1978) Beneficiation of natural zeolites from Bowie, Arizona: a preliminary report: In, Sand, L.B. and Mumpton, F.A., Eds., Natural Zeolites: Occurrence, Properties, Use, Pergamon Press, Elmsford, N.Y., 527-537.

Mori, Y., Edo, Y., Toryu, H., and Ito, T. (1974) Effect of the particle size and application rate of carrier on the herbicidal effect and the growth rate of dry-seeded rice: Zasso Ken Kyu 18, 21-26.

Mumpton, F.A. (1978) Natural zeolites: a new industrial mineral commodity: In, Sand, L.B. and Mumpton, F.A., Eds., Natural Zeolites: Occurrence, Properties, Use, Pergamon Press, Elmsford, N.Y., 1-27.

Mumpton, F.A. and Fishman, P.H. (1977) The application of natural zeolites in animal science and aquaculture: J. Animal Sci. 4, 1188-1203.

Murphy, C.B., Hrycyk, O. and Gleason, W.T. (1977) Single P/C unit removal of nutrients from combined sewer overflows: J. Water Poll. Cont. Fed. 49 (in press).

Neel, E.E.A. et al. (1973) Catalytic manufacture of gasoline for automobile engines: Ger. Offen. 2,259,794, June 14, 25 pp.

Nelson, J.L. and Mercer, B.W. (1963) Ion exchange separation of cesium from alkaline waste supernatant solutions: U.S. Atomic Energy Comm. Doc. HY-76449.

Nemecz, E. and Varju, G. (1962) Sodium bentonization, clinoptylolitization, and ardularization in the rhyolitic tuffs of the Szerencs Piedmont area: Acta Geol. 6, 389-426.

Nichols, R. (1976) Jungle Laboratories, Comfort, Texas (personal communication).

Nikashina, V.A., Zaborskaya, Ye. Yu., Mahalov, Ye. M., and Rubinshteen, R.N. (1974) Selective sorption of strontium by natural clinoptilolite from aqueous solutions: Radiokhim 16, 753-756.

Nikolova, R. and Ivanov, D. (1975) Possible use of natural clinoptilolite as a catalyst carrier: Khim. Ind. (Sofia) 47, 175-177.

Nishita, H. and Haug, R.M. (1972) Influences of clinoptilolite on Sr^{90} and Cs^{137} uptakes by plants: Soil Sci. 114, 149-157.

NRG (1975) NuFuel, a new source of Energy: NRG NuFuel Company Brochure, 6 pp.

Ohtani, S. et al. (1972a) Conversion catalyst for hydrocarbons: Japan. Kokai 72,046,-667, Nov., 5 pp.

Ohtani, S. et al. (1972b) Conversion catalyst for hydrocarbons: Japan. Kokai 72,046-668, Nov., 6 pp.

Onogi, T. (1966) Treating experiments of chicken droppings with zeolitic tuff powder. 2. Experimental use of zeolite-tuffs as dietary supplements for chickens: Rep. Yamagata Stock Raising Inst., 7-18.

Ousterhout, L.E. (1970) Nutritional effects of clays in feed: Feedstuffs 42, 34-36.

Papp, J., Kallo, D., and Schay, G. (1971) Hydromethylation of toluene on clinoptilolite: J. Catalysis 23, 168-182.

Peters, M.D. and Bose, R.J. (1975) Clinoptilolite--a physio-chemical approach to ammonia removal in hatchery and aquaculture water reuse systems: Fish. and Marine Serv. Tech. Rep. 535, 12 pp.

Rohrer, D.M. (1976) Los Alamos Scientific Laboratory, Los Alamos, New Mexico (personal communication).

Roux, A., Huang, A.A., Ma, Y.H., and Swiebel, I. (1973) Sulfur dioxide adsorption on mordenites: Am. Inst. Chem. Eng. Symp. Ser. 134, Gas Purification by Adsorption, 46-53.

Sato, I. (1975) Adsorption of heavy metal ions on zeolite tuff from the Dirika area, Imagane-cho, Oshima Province, Hokkaido-Fundamental Experiment: Chika Shigen Chosajo Hokoku (Hokkaido) 47, 63-66.

Sato, M. and Fukagawa, K. (1976) Treatment for ammoniacal nitrogen-containing water: Japan. Kokai 76,068,967, June 15, 4 pp.

Scherzer, J., Vaughan, D.E.W., and Albers, E.W. (1975) Hydrocarbon cracking catalysts with promoter mixtures: U.S. Patent 3,894,940, April 15, 4 pp.

Semmons, M.J. and Seyfarth, M. (1978) The selectivity of clinoptilolite for certain heavy metals: In, Sand, L.B. and Mumpton, F.A., Eds., Natural Zeolites: Occurrence, Properties, Use, Pergamon Press, Elmsford, N.Y., 517-526.

Slone, W.J., Turner, P.R., and Jester, D.B. (1975) A new closed vertical raceway fish culture system containing clinoptilolite as an ammonia stripper: Proc. 54th Conf. West. Assoc. Game, Fish, Conserv. Commissioners, July 1974, Albuquerque, New Mexico, 381-394.

Smola, V.I., Anusov, S.A., Zihkovskii, V.A., Ket'tsev, N.V., and Torocheshnikov, N.S. (1975) Use of natural zeolites for removing SO_2 from industrial waste gases: Prom. i. San. Ochistka Gazov. Nauch.-tekhn. Sb. 5, 12-14.

Spiridonova, I.A., Torochesnikov, N.S, and Bobylev, V.N. (1975) Study of the properties of ammonium nitrate in the presence of inorganic additives: Tr. Mosk. Khim.-Tekhnol. Inst. 85, 8-9.

Stojanović, D. (1972) Zeolite-containing volcanic tuffs and sedimentary rocks in Serbia: Proc. Serb. Geol. Soc. for 1968-70, 9-20.

Takasaka, A. (1975) Properties and use of the Itaya zeolite from Japan: Eunsai 20, 127-134, 142.

Tamura, K. (1974) High strength porous glass: Japan. Kokai 74,098,817, Sept. 18, 3 pp.

Tamura, T. (1970) Oxygen concentration process: U.S. Patent 3,533,221, Oct. 13, 10 pp.

Tamura, T. (1971) Gas adsorption properties and industrial applications of Japanese tuff: Seminar on Occurrence and Mineralogy of Sedimentary Zeolites in the Circum-Pacific Region; Nikko, Japan; U.S.-Japan Cooperative Science Program (unpublished abstract).

Tchernev, D.I. (1978) Solar energy applications of natural zeolites: In, Sand, L.B. and Mumpton, F.A., Eds., Natural Zeolites: Occurrence, Properties, Use, Pergamon Press, Elmsford, N.Y., 479-485.

Terui, A. et al. (1974) Adsorbing sulfur compounds, dust, soot from flue gas on zeolitic kieselguhr: Japan. Kokai 74,036,580, April 14, 2 pp.

Torii, K. (1974) Utilization of sedimentary zeolites in Japan: Seminar on the Occurrence, Origin, and Utilization of Sedimentary Zeolites in the Circum-Pacific Region; U.S.-Japan Cooperative Science Program, Menlo Park, California (unpublished abstract).

Torii, K., Hotta, M., Onodera, Y., and Asaka, M. (1973) Adsorption of nitrogen-oxygen on zeolite tuff: J. Chem. Soc. Japan 2, 225-232.

Torii, K., Yoshio, O., Makoto, A., and Hiroshi, I. (1971) Character of gas adsorption and separation in mordenite tuff: Kogyo Kagaku Zasshi 74, 2018-2024.

Tsitsishvili, G.V., Urotadze, S.L., Lukin, V.D., and Bagirov, R.M. (1976) Drying and treatment of natural gas by clinoptilolite in an experimental pilot plant: Soobshch. Akad. Nauk Gruz. S.S.R. 81, 369-371.

Tsitsishvili, G.V., Andronikashvili, T.G., Sabelashvili, Sh.D., and Koridzy, Z.I. (1972) Gas chromatographic separation of Ar-O-N on clinosorb: Zak. Rad. 38, 57-58.

Turner, F.T. (1975) Texas Agricultural Experiment Station, Texas A&M Univ., Beaumont, Texas (personal communication).

Ulisch, G. (1975) Inorganic porous materials: Ger. Offen. 2,334,224, Jan. 23, 11 pp.

Union Carbide Corporation (1962) Linde molecular sieve type AW-500: Linde Molecular Sieves Bull. F-1617, 4 pp.

Union Carbide Corporation (1975) Modular UNOX Systems: Sales Brochure, Environmental Systems Dept. F-3577A, 8 pp.

Utada, M., Minato, H., Ishikawa, T., and Yoshizaki, Y. (1974) The alteration zones surrounding Kuroko-type deposits in Nishi-Aizu District, Fukushima Prefecture, with emphasis on the analcime zone as an indicator in exploration for ore deposits: Min. Geol. Spec. Issue 6, 291-302.

Vdovina, E.D., Radyuk, R.I., and Sultanov, A.S. (1976) Use of Uzbekistan natural zeolites for the purification of low-level wastewaters. I. Sorption of radioactive cesium: Radiokhimiya 18, 422-423.

Watanabe, S., Kanaka, Y., and Kuroda, A. (1971) Report on the experimental use of zeolite-tuff as dietary supplements for cattle: Rep. Okayama Prefecture Fed. Agricul. Coop. Assoc., April, 18 pp.

Wearout, J.D. (1971) Purification of gases by adsorption: U.S. Patent 3,594,983, Jan. 1, 27 pp.

White, J.L. and Ohlroggi, A.J. (1974) Ion exchange materials to increase consumption of non-protein nitrogen in ruminants: Canadian Patent 939186, Jan. 2, 30 pp.

Wilding, M.W. and Rhodes, D.W. (1963) Removal of radioisotopes from solution by earth materials from eastern Idaho: U.S. Atomic Energy Comm. Doc. IDO-14624.

Wilding, M.W. and Rhodes, D.W. (1965) Decontamination of radioactive effluent with clinoptilolite: U.S. Atomic Energy Comm. Doc. IDO-14657.

Wilson, T.E. (1975) Greeley and Hanson Engineers, Chicago, Illinois (personal communication).

Yoshida, K. and Utada, M. (1969) A study on alteration of Miocene Green Tuffs in the Kuroko-type mineralization area in Odate Basin, Akita Prefecture: Min. Geol. 18, 332-342.

Yoshida, H., Kurata, A., and Sanga, S. (1976) Removal of heavy metal ions from wastewater using zeolites: Mizu Shori Gijutsu 17, 219-226.

Yoshinaga, E. et al. (1973) Organophosphate-containing agricultural and horticultural granule formation: U.S. Patent 3,708,573.

Appendix I

REPRESENTATIONS AND MODELS OF ZEOLITE CRYSTAL STRUCTURES

Several kinds of models can be used to illustrate the structure of zeolites. Each is described below.

1. Solid Tetrahedra Models

The regular solid tetrahedron is used to represent the geometric arrangement of oxygens in the primary TO_4 building unit in zeolite structures (Figure 1b). Semiregular Archimedean solids are used to represent the idealized geometric arrangement of tetrahedra in the polyhedral building units, such as cube (D4R), hexagonal prism (D6R), and truncated octahedron (sodalite unit) (Figure 2).

2. Framework Models

The TO_4 tetrahedron can be portrayed by a "jack-like" representation of the four tetrahedral M-O linkages (Figure 1c). Actual models make use of metal or plastic "jacks," usually linked by plastic tubing. The center of the plastic tubing connector represents the position of the oxygen atom. This is the simplest and fastest way to build models of tetrahedral frameworks.

3. Space-Filling Models

These models are drawn or constructed to represent the true volume of oxygen ions and the packing of spheres of oxygen ions in the structure (Figure 1d). It should be remembered that in ionic oxides, e.g., aluminosilicate zeolites, the oxygen ions occupy about 90% of the atomic volume. Therefore, the oxygen-packing or space-filling models are the most realistic view of the structure, but the models are much more difficult to build. A packing drawing of the structure of Na_2CaSiO_4 is shown in Figure 3.

4. Ball and Stick Models

These models are used generally to represent the spatial arrangements of atoms in crystals. The ball represents the atom, the stick the bonding between atoms. Because of the large number of atoms in the structure of zeolites, this method is tedious and impractical.

205

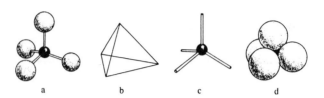

Figure 1.   Methods of representing the tetrahedral coordination of oxygen ions with aluminum and silicon by means of (a) ball and stick model, (b) solid tetrahedron, (c) skeletal tetrahedron, and (d) a space-filling model based on packed spheres (From Breck, 1974, Fig. 2.2.)

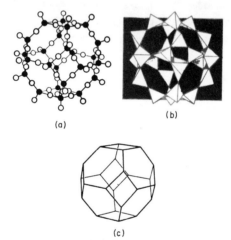

(a)

(b)

(c)

Figure 2.   Three ways of depicting the truncated octahedron (or sodalite unit) in aluminosilicate frameworks. (From Smith, 1976.)

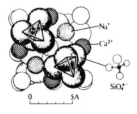

Figure 3.   A packing drawing of the structure of Na_2CaSiO_4. The location of Na^+ and Ca^{+2} ions between individual SiO_4^{-4} ions is indicated. (From Breck, 1974, Fig. 2.3.)

REFERENCES: See Chapter 2. p. 52.

Appendix II

CRYSTAL STRUCTURE DATA FOR IMPORTANT ZEOLITES

Tables of crystal structure data from D.W. Breck (1974) *Zeolite Molecular Sieves* (Wiley-Interscience, New York, 771 pp) are reproduced here with the permission of the author and Wiley-Interscience.

Table 1. Analcime, Table 2.20, p. 135, Breck

Structure Group:	1
Reference:	62, 67, 173–176
Chemical Composition	
Typical Oxide Formula:	$Na_2O \cdot Al_2O_3 \cdot 4\, SiO_2 \cdot 2\, H_2O$
Typical Unit Cell Contents:	$Na_{16}[(AlO_2)_{16}(SiO_2)_{32}] \cdot 16\, H_2O$
Variations:	$Si/Al = 1.8–2.8;\ H_2O = 14–18$
Crystallographic Data	
Symmetry:	Cubic Density: 2.25 g/cc
Space Group:	Ia3d Unit Cell Volume: 2590 A^3
Unit Cell Constants:	$a = 13.72$ A X-Ray Powder Data: Table 3.4
Structural Properties	
Framework: Stereo 2.4	S4R of type UDUD connected to form regular S6R which lie ‖ to (111). Also very distorted 8-rings.
SBU:	S4R Void volume: 0.18 cc/cc
Cage type:	None specific Framework density: 1.85 g/cc
Channel System: Fig. 2.26	One-dimensional, ‖ to [111]
Hydrated—	
Free Apertures:	6-ring, 2.6 A
Cation Locations:	16 in 24 sites; in distorted octahedra with 4 cations and 2 H_2O molecules
Dehydrated—	
Free Apertures:	Unknown
Cation Locations:	Unknown
Effect of Dehydration:	Continuous and reversible; stable to 700°C
Location of H_2O Molecules:	On 3-fold axes in the channels
Largest Molecule Adsorbed:	NH_3
Kinetic Diameter, σ, A:	2.6

Table 2. Wairakite, Table 3.35, p. 240, Breck

Structure Group:	1
Typical Unit Cell Contents:	$Ca_8[(AlO_2)_{16}(SiO_2)_{32}] \cdot 16\ H_2O$
Occurrence:	New Zealand, California, Washington; sedimentary
System:	Monoclinic, pseudocubic, a = 13.69, b = 13.68
	c = 13.56. β = 90.5°
Habit:	Microscopic granules
Twinning:	[110]
Density:	2.265
Hardness:	5½ - 6
Optical Properties:	Biaxial (-), α = 1.498, γ = 1.502; birefringence,
	δ = 0.004, 2Vγ = 70 - 105°
Reference:	9, 40

X-Ray Powder Data (40)

hkl [a]	d(A)	I	hkl	d(A)	I	hkl	d(A)	I
(200)	6.85	40	440	2.418	30	633,	1.857	30
211	5.57	80	(530),	2.35	<10B	721,		
220	4.84	40	(433)			552		
321	3.64	30	(600),	2.26-2.28	10B	–	1.844	10
400	3.42	60	(422)			642	1.822	<10B
	3.39	100	611,	2.215	40	732,	1.722-1.732	40B
(411),	3.21	<10B	532			651		
(330)			620	2.17	<10	800	1.708	<10
420	3.04-3.06	10B		2.147	10		1.696	<10
322	2.909	50	541	2.115	10	741	1.680	20B
	2.897	30		2.095	<10	820,	1.66	<10B
422	2.783	10	631	1.996	20	(644)		
	2.770	10	543	1.93	<10B	822,	1.612	10B
431	2.680	40	640	1.886-1.895	30B	660		
(510)	2.67	10	–	1.867	10	831	1.595	<10
521	2.50	<10					1.586	20
	2.489	40						

[a] Indices in parentheses are based on pseudocubic unit cell.
B = broad

Table 3. Phillipsite, Table 2.56, p. 170, Breck

Structure Group:	1
Other Designation:	Wellsite
Reference:	74, 208

Chemical Composition

Typical Oxide Formula:	$(Ca,Na_2,K_2)O \cdot Al_2O_3 \cdot 4.4\ SiO_2 \cdot 4\ H_2O$
Typical Unit Cell Contents:	$(Ca,K_2,Na_2)_5 [(AlO_2)_{10}(SiO_2)_{22}] \cdot 20\ H_2O$
Variations:	Si/Al = 1.7 to 2.4; Ba found in wellsite; K found in all cases

Crystallographic Data

Symmetry:	Orthorhombic	Density:	2.15 g/cc
Space Group:	B2mb	Unit Cell Volume:	2022 A³
Unit Cell Constants:	a = 9.96 A	X-Ray Powder Data:	Table 3.30
	b = 14.25 A		
	c = 14.25 A		

Structural Properties

Framework:	Stereo 2.5	4-rings of type UUDD crosslinked so that chains of UUDD units run ‖ to a axis
	SBU:	S4R Void volume: 0.31 cc/cc
		Framework density: 1.58 g/cc
Channel System:		Three-dimensional; ‖ to a, ‖ to b, ‖ to c

Hydrated–

Free Apertures:	8-rings, 4.2 x 4.4 A ⊥ to a, 2.8 x 4.8 A ⊥ to b, 3.3 A ⊥ to c
Cation Locations:	In channels, coordinated to H_2O and framework

Dehydrated–

Free Apertures:	Unknown
Effect of Dehydration:	Structure degrades at 200°C
Location of H_2O Molecules:	Coordinated to cations
Largest Molecule Adsorbed:	H_2O
Kinetic Diameter, σ, A:	2.6

Table 4. Harmotome, Table 2.38, p. 153, Breck

Structure Group:	1
Reference:	73, 195
Chemical Composition	
Typical Oxide Formula:	$BaO \cdot Al_2O_3 \cdot 6 SiO_2 \cdot 6 H_2O$
Typical Unit Cell Contents:	$Ba_2[(AlO_2)_4(SiO_2)_{12}] \cdot 12 H_2O$
Variations:	$Ba \gg K$ or Na; Si/Al = 2.3–2.5
Crystallographic Data	
Symmetry:	Monoclinic Density: 2.35 g/cc
Space Group:	$P2_1$ Unit Cell Volume: 999 A^3
Unit Cell Constants:	a = 9.87 A X-Ray Powder Data: Table 3.19
	b = 14.14 A
	c = 9.72 A
	$\beta = 124°50'$

Structural Properties		
Framework: Stereo 2.5	Same as phillipsite	
SBU:	S4R	Void volume: 0.31 cc/cc
		Framework density: 1.59 g/cc
Channel System:	Three-dimensional; ‖ to a, ‖ to b, ‖ to c	
Hydrated—		
Free Apertures:	8-rings, 4.2 x 4.4, 2.8 x 4.8 A, and 3.3 A	
Cation Locations:	Coordinated to 6 O's, 4 H_2O	
Dehydrated—		
Free Apertures:	Unknown	
Effect of Dehydration:	Stable	
Location of H_2O Molecules:	8-coordinated to Ba^{2+}	
Largest Molecule Adsorbed:	NH_3	
Kinetic Diameter, σ, A:	2.6	

Table 5. Gismondine, Table 2.34, Breck

Structure Group:	1
Reference:	75, 76
Chemical Composition	
Typical Oxide Formula:	$CaO \cdot Al_2O_3 \cdot 2 SiO_2 \cdot 4 H_2O$
Typical Unit Cell Contents:	$Ca_4[(AlO_2)_8(SiO_2)_8] \cdot 16 H_2O$
Variations:	Si/Al = 1.12–1.49; some K
Crystallographic Data	
Symmetry:	Monoclinic Density: 2.27 g/cc
Space Group:	$P2_1/C$ Unit Cell Volume: 1046 A^3
Unit Cell Constants:	a = 9.84 A X-Ray Powder Data: Table 3.16
	b = 10.02 A
	c = 10.62 A $\beta = 92°25'$

Structural Properties		
Framework: Stereo 2.6	4-rings, crosslinked so that UUDD chains ‖ a and	
	b; Si-Al ordered	
SBU:	S4R	Void volume: 0.46 cc/cc
Cage type:	nonspecific	Framework density: 1.52 g/cc
Channel System: Fig. 2.30b	Three-dimensional; ‖ to a, ‖ to b	
	Tube bundle type	
Hydrated—		
Free Apertures:	8-rings; 3.1 x 4.4 A ⊥ to a, 2.8 x 4.9 A ⊥ to b	
Cation Locations:	At channel intersection coordinated to 4 H_2O	
	and 2 cations.	
Dehydrated—		
Free Apertures:	Unknown	
Cation Locations:	Unknown	
Effect of Dehydration:	Stable to 250°C	
Location of H_2O Molecules:	Coordinated to Ca^{2+}; H_2O/Ca^{2+} = 4	
Largest Molecule Adsorbed:	H_2O	
Kinetic Diameter, σ, A:	2.6	

Table 6. Laumontite, Table 2.43, Breck

Structure Group:	1
Other Designation:	Leonhardite (partially dehydrated)
Reference:	32, 82, 83
Chemical Composition	
Typical Oxide Formula:	$CaO \cdot Al_2O_3 \cdot 4\,SiO_2 \cdot 4\,H_2O$
Typical Unit Cell Contents:	$Ca_4\,[(AlO_2)_8\,(SiO_2)_{16}]\cdot 16\,H_2O$
Variations:	K, Na; Si/Al = 1.75–2.28
Crystallographic Data	

Symmetry:	Monoclinic	Density:	2.30 g/cc
Space Group:	Am	Unit Cell Volume:	1390 A^3
Unit Cell Constants:	a = 14.90 A	X-Ray Powder Data:	Table 3.23
	b = 13.17 A		
	c = 7.50 A		
	$\beta = 111°30'$		

Structural Properties		
Framework: Stereo 2.10	4-rings (Si_4O_{12}) in TTTT configuration ‖ to (120) and linked by AlO_4 tetrahedra	
SBU:	S4R	Void volume: 0.34 cc/cc
		Framework density: 1.77 g/cc
Channel System:	One-dimensional, ‖ to a; 4.6 x 6.3 A	
Hydrated—		
Free Apertures:	10-ring, distorted 4.6 x 6.3 A	
Cation Locations:	4 O and 2 H_2O in 6-fold coordination	
Dehydrated—		
Free Apertures:		
Cation Locations:		
Effect of Dehydration:	Stepwise dehydration	
Largest Molecule Adsorbed:	H_2O, NH_3 if dehydrated at 200°	
Kinetic Diameter, σ, A:	2.6	

Table 7. Erionite, Table 2.28, Breck

Structure Group:	2
Reference:	88, 89, 90, 91, 92
Chemical Composition	
Typical Oxide Formula:	$(Ca,Mg,Na_2,K_2)O \cdot Al_2O_3 \cdot 6\,SiO_2 \cdot 6\,H_2O$
Typical Unit Cell Contents:	$(Ca,Mg,Na_2,K_2)_{4.5}\,[(AlO_2)_9\,(SiO_2)_{27}]\cdot 27\,H_2O$
Variations:	Si/Al = 3–3.5. Alkali ions > alkaline earths. Possibly Fe^3 in tetrahedral sites.
Crystallographic Data	

Symmetry:	Hexagonal	Density:	2.02 g/cc
Space Group:	$P6_3/mmc$	Unit Cell Volume:	2300 A^3
Unit Cell Constants:	a = 13.26 A	X-Ray Powder Data:	Table 3.12
	c ≐ 15.12 A		

Structural Properties		
Framework: Stereo 2.12	Units of S6R arranged ‖ in stacking sequence of AABAAC. Columns of D6R and ϵ-cages in c direction in sequence ϵ-D6R-ϵ linked by 6-rings	
SBU:	S6R and D6R	Void volume: 0.35 cc/cc
Cage type:	ϵ, 23-hedron	Framework density: 1.51 g/cc
Channel System: Fig. 2.29c	Three-dimensional, ⊥ to c one-dimensional, 2.5 A, ‖ to c	
Hydrated—		
Free Apertures:	8-ring, 3.6 x 5.2 A	
Cation Locations:	In ϵ-cages	
Dehydrated—		
Effect of Dehydration:	Stable	
Largest Molecule Adsorbed:	n-paraffin hydrocarbons	
Kinetic Diameter, σ, A:	4.3	

210

Table 8.   Offretite, Table 2.52, Breck

Structure Group:	2
Reference:	90, 91, 93, 192, 204
Chemical Composition	
Typical Oxide Formula:	0.28 MgO, 0.41 CaO, 0.26 $K_2O \cdot Al_2O_3 \cdot$
	4.98 $SiO_2 \cdot 5.9\ H_2O$
Typical Unit Cell Contents:	$(K_2,Ca,Mg)_{2.5}[(AlO_2)_5(SiO_2)_{13}] \cdot 15\ H_2O$
Variations:	Alkaline earths > Alkali metals
Crystallographic Data	
Symmetry:	Hexagonal Density: 2.13 g/cc
Space Group:	P6m2 Unit Cell Volume: 1160 A^3
Unit Cell Constants:	a = 13.291 ± .002 X-Ray Powder Data: Table 3.28
	c = 7.582 ± .006
Structural Properties	
Framework: Stereo 2.13	Units of S6R arranged ‖ to c in sequence
	AABAAB ... Columns of ϵ-cages and D6R in c-
	direction
SBU:	D6R, S6R Void volume: 0.40 cc/cc
Cage Type	ϵ, and 14-hedron II Framework density: 1.55 cc/cc
Channel System: Fig. 2.29b	‖ to c, 6.4 A; ‖ to a, 3.6 x 5.2 A
Hydrated--	
Free Apertures:	3.6 x 5.2 A; 6.4 A
Cation Locations:	K^+ in ϵ-cages, and not exchangeable;
	Ca^{2+} is in D6R; Mg^{2+} is in the 14-hedron cage
Dehydrated—	
Free Apertures:	Unknown
Location of H_2O Molecules:	10 are coordinated to cations; 5 mobile in large
	channels
Largest Molecule Adsorbed:	Cyclohexane
Kinetic Diameter, σ, A:	6

Table 9.   Hydrated Sodalite, Table 2.40, Breck

Structure Group:	2
Other Designation:	Sodalite Hydrate, Hydroxy Sodalite, G
Reference:	99. 100, 196
Chemical Composition	
Typical Oxide Formula:	$Na_2O \cdot Al_2O_3 \cdot 2\ SiO_2 \cdot 2.5\ H_2O$
Typical Unit Cell Contents:	$Na_6[(AlO_2)_6(SiO_2)_6]{\sim}7.5\ H_2O$
Variations:	May contain varying amounts of NaOH
Crystallographic Data	
Symmetry:	Cubic Density: 2.03 g/cc
Space Group:	P$\bar{4}$3n Unit Cell Volume: 702 A^3
Unit Cell Constants:	a = 8.86 A X-Ray Powder Data: Table 4.28
Structural Properties	
Framework: Stereo 2.22	Units of S6R ‖ and connected in sequence
	ABCABC. Close packed truncated octahedral
	cages or β-cages
SBU:	S6R Void volume: 0.35 cc/cc
Cage type:	β Framework density: 1.72 g/cc
Channel System:	Three-dimensional, interconnected β-cages
Hydrated—	
Free Apertures:	6-rings, 2.2 A
Cation Locations:	Near the 6-rings but in the cavity of the β-cage
Dehydrated—	
Cation Locations:	Probably on the 6-rings
Effect of Dehydration:	None to 800°C
Location of H_2O Molecules:	Probably 4 in each β-cage
Largest Molecule Adsorbed:	H_2O
Kinetic Diameter, σ, A:	2.6

Table 10. Zeolite A, Table 2.18, Breck

Structure Group:	3
Reference:	102, 103, 105–112, 115, 172

Chemical Composition

Typical Oxide Formula:	$Na_2O \cdot Al_2O_3 \cdot 2\ SiO_2 \cdot 4.5\ H_2O$
Typical Unit Cell Contents:	$Na_{12}\,[(AlO_2)_{12}(SiO_2)_{12}] \cdot 27\ H_2O$, pseudo cell and 8X for true cell
Variations:	$Si/Al = {\sim}0.7$ to 1.2; occlusion of $NaAlO_2$ in β-cages

Crystallographic Data

Symmetry:	Cubic	Density:	1.99 g/cc
Space Group:	Pm3m	Unit Cell Volume:	1870 A^3
	(Fm3c for true cell)		pseudo cell
Unit Cell Constants:	a = 12.32 A, pseudo cell	X-Ray Powder Data:	Table 4.26
	a = 24.64 A for true cell		

Structural Properties

Framework:	Stereo 2.16	Cubic array of β-cages linked by D4R units
	SBU: D4R	Void volume: 0.47 cc/cc
	Cage type: α, β	Framework density: 1.27 g/cc
	(one each)	
Channel System:	Fig. 2.28a	Three-dimensional, $\parallel$ to [100]; 4.2 A and $\parallel$ to [111]; 2.2 A minimum diameter

Hydrated—

Free Apertures:	2.2 A into β-cage and 4.2 A into α-cage
Cation Locations:	8 S_I on 6-rings, 4 cations with H_2O in the 8-rings

Dehydrated—

Free Apertures:	4.2 A
Cation Locations:	8 S_I in 6-rings, 3 S_{II} in 8-rings, 1 S_{III} at the 4-ring
Effect of Dehydration:	None on framework, 4 cations move to S_{II}
Location of H_2O Molecules:	Dodecahedral arrangement in α-cage 4 molecules in β-cage.
Largest Molecule Adsorbed:	C_2H_4 at RT, O_2 at $-183°C$
Kinetic Diameter, σ, A:	3.9 and 3.6

Table 11. Faujasite, Table 2.30, Breck

Structure Group:	4
Reference:	59, 123, 125, 131–133, 187–190

Chemical Composition

Typical Oxide Formula:	$(Na_2,Ca,Mg,K_2)O \cdot Al_2O_3 \cdot 4.5\ SiO_2 \cdot 7\ H_2O$
Typical Unit Cell Contents:	$Na_{12}Ca_{12}Mg_{11}\,[(AlO_2)_{59}(SiO_2)_{133}] \cdot 235\ H_2O$
Variations:	K observed in variable amounts Mg observed in variable amounts

Crystallographic Data

Symmetry:	Cubic	Density:	1.91 g/cc
Space Group:	Fd3m	Unit Cell Volume:	15,014 A^3
Unit Cell Constants:	a = 24.67 A	X-Ray Powder Data:	Table 3.13

Structural Properties

Framework:	Stereo 2.17	Truncated octahedra β-cages, linked tetrahedrally through D6R's in arrangement like carbon atoms in diamond. Contains eight cavities $\sim$ 13 A in diameter in each unit cell.
	SBU:	D6R, 16/uc Void volume: 0.47 cc/cc
	Cage type:	β, 8/uc, 26-hedron (II). Framework density: 1.27g/cc
Channel System:	Fig. 2.30	Three-dimensional, $\parallel$ to [110]

Hydrated—

Free Apertures:	12-ring 7.4 A; 6-ring 2.2 A
Cation Locations:	See Table 2.11

Dehydrated—

Cation Locations:	See Table 2.11
Effect of Dehydration:	Stable and reversible
Location of H_2O Molecules:	4 in each β-cage
Largest Molecule Adsorbed:	$(C_2F_5)_3N$
Kinetic Diameter, σ, A:	8.0

Table 12. Zeolite X, Table 2.63, Breck

Structure Group:	4
Reference:	59, 78, 104, 107, 124, 138, 214, 215
Chemical Composition	
Typical Oxide Formula:	$Na_2O \cdot Al_2O_3 \cdot 2.5\ SiO_2 \cdot 6\ H_2O$
Typical Unit Cell Contents:	$Na_{86} [(AlO_2)_{86}(SiO_2)_{106}] \cdot 264\ H_2O$
Variations:	Ga substitution for Al; Si/Al = 1 to 1.5
	Na/Al = 0.7 to 1.1

Crystallographic Data

Symmetry:	Cubic	Density:	1.93 g/cc
Space Group:	Fd3m	Unit Cell Volume: 15,362–	
Unit Cell Constants:	a = 25.02–24.86 A		15,670 A^3
		X-Ray Powder Data: Table 4.88	

Structural Properties

Framework:	Stereo 2.17	Truncated octahedra, β-cages, linked tetrahedrally through D6R's in arrangement like carbon atoms in diamond. Contains eight cavities, ~ 13 A in diameter in each unit cell
	SBU:	D6R, Void volume: 0.50 cc/cc
	Cage type:	β, 26-hedron (II) Framework density: 1.31 g/cc
Channel System: Fig. 2.30		Three-dimensional, ‖ to [110]
Hydrated–		
Free Apertures:		12-ring, 7.4 A, 6-ring, 2.2 A
Cation Locations:		Table 2.12
Dehydrated–		
Free Apertures:		7.4 A
Cation Locations:		Table 2.12
Effect of Dehydration:		Stable and reversible
Location of H$_2$O Molecules:		See Table 2.12
Largest Molecule Adsorbed:		$(C_4H_9)_3N$
Kinetic Diameter, σ, A:		8.1

Table 13. Zeolite Y, Table 2.64, Breck

Structure Group:	4
Reference:	59, 124, 128, 140, 216
Chemical Composition	
Typical Oxide Formula:	$Na_2O \cdot Al_2O_3 \cdot 4.8\ SiO_2 \cdot 8.9\ H_2O$
Typical Unit Cell Contents:	$Na_{56} [(AlO_2)_{56}(SiO_2)_{136}] \cdot 250\ H_2O$
Variations:	Na/Al 0.7 to 1.1; Si/Al => 1.5 to about 3
Crystallographic Data	

Symmetry:	Cubic	Density:	1.92 g/cc
Space Group:	Fd3m	Unit Cell Volume: 14,901 to	
Unit Cell Constants:	a = 24.85–24.61 A		15,347 A^3
		X-Ray Powder Data: Table 4.90	

Structural Properties

Framework:	Stereo 2.17	Truncated octahedra, β-cages, linked tetrahedrally through D6R's in arrangement like carbon atoms in diamond. Contains eight cavities ~ 13 A in diameter in each unit cell.
	SBU:	D6R Void volume: 0.48 cc/cc
	Cage type:	β, 26-hedron (II) Framework density: 1.25–1.29 g/cc
Channel System: Fig. 2.30		Three-dimensional, ‖ to [110]
Hydrated–		
Free Apertures:		12-ring, 7.4 A; 6-ring, 2.2 A
Cation Locations:		Table 2.13
Dehydrated–		
Free Apertures:		~ 7.4
Cation Locations:		Table 2.13
Effect of Dehydration:		Stable and reversible
Location of H$_2$O Molecules:		Not specifically located
Largest Molecule Adsorbed:		$(C_4H_9)_3N$
Kinetic Diameter, σ, A:		8.1

Table 14. Chabazite, Table 2.23, Breck

Structure Group:	4
Reference:	143, 150, 152, 178

Chemical Composition

Typical Oxide Formula:	$CaO \cdot Al_2O_3 \cdot 4\ SiO_2 \cdot 6.5\ H_2O$
Typical Unit Cell Contents:	$Ca_2\ [(AlO_2)_4(SiO_2)_8] \cdot 13\ H_2O$
Variations:	Na,K; Si/Al = 1.6–3

Crystallographic Data

Symmetry:	Rhombohedral	Density: 2.05–2.10 g/cc
Space Group:	$R\bar{3}m$	Unit Cell Volume: 822 A^3
Unit Cell Constants:	a = 9.42 A	X-Ray Powder Data: Table 3.7
	$\beta = 94°28'$	
Hexagonal,	a = 13.78; c = 15.06	

Structural Properties

Framework: Stereo 2.19		Configuration of D6R units in sequence ABCABC linked by tilted 4-rings
	SBU:	D6R Void volume: 0.47 cc/cc
	Cage type:	Ellipsoidal, Framework density: 1.45 g/cc
		6.7 x 10 A
Channel System: Fig. 2.29a	Three-dimensional	

Hydrated–

Free Apertures:	8-rings, 3.7 x 4.2 A
	6-rings, 2.6 A
Cation Locations:	Coordinated with 4 H_2O in the cavity

Dehydrated–

Free Apertures:	3.1 x 4.4 A
Cation Locations:	0.6 Ca^{2+} in S_I; 0.35 Ca^{2+} in S_{II}; 1/16 Ca^{2+} in S_{III}
Effect of Dehydration:	Some framework distortion
Location of H_2O Molecules:	In cavities, 5 per Ca^{2+} ion
Largest Molecule Adsorbed:	*n*-paraffin hydrocarbons
Kinetic Diameter, σ, A:	4.3

Table 15. Natrolite, Table 2.50, Breck

Structure Group:	5
Other Designation:	Elagite
Reference:	158, 198, 200, 201

Chemical Composition

Typical Oxide Formula:	$Na_2O \cdot Al_2O_3 \cdot 3\ SiO_2 \cdot 2\ H_2O$
Typical Unit Cell Contents:	$Na_{16}\ [(AlO_2)_{16}(SiO_2)_{24}] \cdot 16\ H_2O$
Variations:	Si/Al = 1.44–1.58; H_2O/Na = 1; Ca, K very small

Crystallographic Data

Symmetry:	Orthorhombic	Density: 2.23 g/cc
Space Group:	Fdd2	Unit Cell Volume: 2250 A^3
Unit Cell Constants:	a = 18.30 A	X-Ray Powder Data: Table 3.27
	b = 18.63 A	
	c = 6.60 A	

Structural Properties

Framework: Stereo 2.24b		Crosslinked chains of 4-1 units. Al and Si atoms are ordered.
	SBU:	Unit of 4-1 Void volume: 0.23 cc/cc
		4-1 $[Al_2Si_3O_{10}]$ Framework density: 1.76 g/cc
Channel System: Fig. 2.27	Two-dimensional ⊥ to c	

Hydrated–

Free Apertures:	8-rings, 2.6 x 3.9 A
Cation Locations:	In channels coordinated to 2 H_2O and 4 framework O atoms

Dehydrated–

Free Apertures:	Unknown
Cation Locations:	Unknown
Effect of Dehydration:	Framework shrinks due to rotation of chains
Location of H_2O Molecules:	In channels coordinated to oxygen in framework and sodium ions
Largest Molecule Adsorbed:	NH_3
Kinetic Diameter, σ, A:	2.6

Table 16.   Thomsonite, Table 2.60, Breck

Structure Group:	5
Other Designation:	Faroeite
Reference:	183, 194, 198, 212
Chemical Composition	
Typical Oxide Formula:	$(Na_2, Ca)O \cdot Al_2O_3 \cdot 2\ SiO_2 \cdot 2.4\ H_2O$
Typical Unit Cell Contents:	$Na_4Ca_8\ [(AlO_2)_{20}(SiO_2)_{20}] \cdot 24\ H_2O$
Variations:	Si/Al = 1.0–1.1; Gonnardite, Si/Al = 1.5
	Faroeite is Na/Ca = 1
Crystallographic Data	
Symmetry:	Orthorhombic Density: 2.3 g/cc
Space Group:	Pnn2 Unit Cell Volume: 2253 A^3
Unit Cell Constants:	a = 13.07 A X-Ray Powder Data: Table 3.33
	b = 13.08 A
	c = 13.18 A
Structural Properties	
Framework: Stereo 2.25	Crosslinked chains of 4-1 units Si/Al = 1 and Al, Si are ordered
SBU:	Unit of 4-1 Void volume: 0.32 cc/cc
	$[Al_{2.5}Si_{2.5}O_{10}]$ Framework density: 1.76 g/cc
Channel System: Fig. 2.27	Two-dimensional and ⊥ to c
Hydrated–	
Free Apertures:	8-ring, 2.6 x 3.9 A
Cation Locations:	8 coordinated with 4 cations and 3 H_2O's
	Remainder in 8-fold coordination
Location of H_2O Molecules:	Zigzag chains in the channels
Largest Molecule Adsorbed:	NH_3
Kinetic Diameter, σ, A:	2.6

Table 17.   Edingtonite, Table 2.26, Breck

Structure Group:	5
Reference:	183–186
Chemical Composition	
Typical Oxide Formula:	$BaO \cdot Al_2O_3 \cdot 3\ SiO_2 \cdot 4\ H_2O$
Typical Unit Cell Contents:	$Ba_2\ [(AlO_2)_4(SiO_2)_6] \cdot 8\ H_2O$
Variations:	May have some Ca, Na, K
Crystallographic Data	
Symmetry:	Orthorhombic Density: 2.75 g/cc
Space Group:	P222 Unit Cell Volume: 598 A^3
Unit Cell Constants:	a = 9.54 A X-Ray Powder Data: Table 3.10
	b = 9.65 A
	c = 6.50 A
Structural Properties	
Framework: Stereo 2.26	Crosslinked chains of 4-1 units, Si/Al = 1.5 Al, Si ordered
SBU:	Unit of 4-1 Void volume: 0.36 cc/cc
	$[Al_2Si_3O_{10}]$ Framework density: 1.68 g/cc
Channel System: Fig. 2.27	Two-dimensional ⊥ to c
Hydrated–	
Free Apertures:	8-ring, 3.5 x 3.9
Cation Locations:	In alternate cavities, 8-coordinated with 6 oxygens and 2 H_2O
Dehydrated–	
Effect of Dehydration:	Stable at 250°C
Location of H_2O Molecules:	Double chains in each channel
Largest Molecule Adsorbed:	H_2O
Kinetic Diameter, σ, A:	2.6

Table 18. Mordenite, Table 2.47, Breck

Structure Group:	6
Other Designation:	Ptilolite, Arduinite, Flokite, Deeckite
Reference:	89, 159, 161

Chemical Composition

Typical Oxide Formula:	$Na_2O \cdot Al_2O_3 \cdot 10\ SiO_2 \cdot 6\ H_2O$
Typical Unit Cell Contents:	$Na_8\ [(AlO_2)_8(SiO_2)_{40}] \cdot 24\ H_2O$
Variations:	Si/Al = 4.17–5.0; Na, Ca > K

Crystallographic Data

Symmetry:	Orthorhombic	Density:	2.13 g/cc
Space Group:	Cmcm	Unit Cell Volume:	2794 A^3
Unit Cell Constants:	a = 18.13	X-Ray Powder Data:	Table 3.26
	b = 20.49		
	c = 7.52		

Structural Properties

Framework: Stereo 2.27	Complex chains of 5-rings crosslinked by 4-rings Chains consist of 5-rings of SiO_4 tetrahedra and single AlO_4 tetrahedra	
SBU:	Unit 5-1	Void volume: 0.28 cc/cc
		Framework density: 1.70 g/cc
Channel System: Fig. 2.27a	Main system ‖ to c; channels with 2.8 A restrictions ‖ to b	

Hydrated—

Free Apertures:	12-rings ⊥ to c-axis, 6.7 x 7.0 A
	8-rings, 2.9 x 5.7 A, ⊥ to b
Cation Locations:	Four Na⁺ in the restrictions with minimum dimension of 2.8 A in channels ⊥ to b

Dehydrated—

Free Apertures:	Probably no change
Cation Locations:	Unknown
Effect of Dehydration:	Very stable
Location of H_2O Molecules:	Unknown
Largest Molecule Adsorbed:	C_2H_4
Kinetic Diameter, σ, A:	3.9

Table 19. Dachiardite, Table 2.25, Breck

Structure Group:	6
Reference:	162, 182

Chemical Composition

Typical Oxide Formula:	$Na_2O \cdot Al_2O_3 \cdot 7.6\ SiO_2 \cdot 4.8\ H_2O$
Typical Unit Cell Contents:	$Na_5\ [(AlO_2)_5(SiO_2)_{19}] \cdot 12\ H_2O$
Variations:	Other cations Na, K, Ca, Mg

Crystallographic Data

Symmetry:	Monoclinic	Density:	2.16 g/cc
Space Group:	C2/m	Unit Cell Volume:	1384 A^3
Unit Cell Constants:	a = 18.73 A	X-Ray Powder Data:	Table 3.9
	b = 7.54 A		
	c = 10.30 A		
	β = 107°54′		

Structural Properties

Framework: Stereo 2.28	Complex chains of 5-rings; crosslinked by 4-rings	
SBU:	Unit of 5-1	Void volume: 0.32 cc/cc
		Framework density: 1.72 g/cc
Channel System: Fig. 2.27	Two-dimensional, ‖ to c-axis and b-axis.	

Hydrated—

Free Apertures:	8-ring, 3.7 x 6.7 A
	10-ring, 3.6 x 4.8 A

Dehydrated—

Effect of Dehydration:	Probably stable

Table 20. Ferrierite, Table 2.31, Breck

Structure Group:	6
Reference:	163, 164, 181
Chemical Composition	
Typical Oxide Formula:	$(Na_2,Mg)O \cdot Al_2O_3 \cdot 11.1\ SiO_2 \cdot 6.5\ H_2O$
Typical Unit Cell Contents:	$Na_{1.5}Mg_2\ [(AlO_2)_{5.5}(SiO_2)_{30.5}] \cdot 18\ H_2O$
Variations:	Si/Al to 3.8; some Fe^{3+}; also K and Ca; H_2O to 23
Crystallographic Data	

Symmetry:	Orthorhombic	Density:	2.13 - 2.14 g/cc
Space Group:	Immm	Unit Cell Volume:	2027 A^3
Unit Cell Constants:	a = 19.16 A	X-Ray Powder Data:	Table 3.14
	b = 14.13 A		
	c = 7.49 A		

Structural Properties		
Framework: Stereo 2.29	Complex chains of 5-rings which ‖ c-axis are cross-linked by 4-rings	
SBU:	Unit of 5-1	Void volume: 0.28 cc/cc
		Framework density: 1.76 g/cc
Channel System: Fig. 2.27	Two-dimensional ‖ to c-axis, and b-axis	
Hydrated—		
Free Apertures:	10-ring, 4.3 x 5.5 A; 8-ring, 3.4 x 4.8 A	
Cation Locations:	Mg^{2+} as $Mg(H_2O)_6{}^{2+}$ ions in center of cavities	
Dehydrated—		
Cation Locations:	Unknown	
Effect of Dehydration:	Stable	
Location of H_2O Molecules:	In hydration sphere of Mg^{2+}	
Largest Molecule Adsorbed:	C_2H_4	
Kinetic Diameter, σ, A:	3.9	

Table 21. Heulandite, Table 2.39, Breck

Structure Group:	7
Reference:	168
Chemical Composition	
Typical Oxide Formula:	$CaO \cdot Al_2O_3 \cdot 7\ SiO_2 \cdot 6\ H_2O$
Typical Unit Cell Contents:	$Ca_4\ [(AlO_2)_8(SiO_2)_{28}] \cdot 24\ H_2O$
Variations:	Also K and Sr; Si/Al 2.47—3.73; water rich variety has 30 H_2O
Crystallographic Data	

Symmetry:	Monoclinic	Density:	2.198 g/cc
Space Group:	Cm	Unit Cell Volume:	2103 A^3
Unit Cell Constants:	a = 17.73 A	X-Ray Powder Data:	Table 3.21
	b = 17.82 A		
	c = 7.43 A		
	$\beta = 116°20'$		

Structural Properties		
Framework: Stereo 2.32	Special configuration of tetrahedra in 4- and 5-rings arranged in sheets ‖ to (010)	
SBU:	Unit 4-4-1	Void volume: 0.39 cc/cc
		Framework density: 1.69 cc/cc
Channel System: Fig. 2.27	Two-dimensional, consisting of three channels, ‖ to a-axis and c-axis and at 50° to a-axis	
Hydrated—		
Free Apertures:	8-ring, 4.0 x 5.5 A ⊥ to a; 10-ring, 4.4 x 7.2 A ⊥ to c; 8-ring, 4.1 x 4.7 A ⊥ to c	
Cation Locations:	Three types located in open channels. Ca_1 and Ca_2 in eight-fold coordination with 3 framework oxygens and 5 water molecules near channel walls. Ca_3 is similarly coordinated and located at intersection of 8-ring channels.	

Continued next page →

Table 21. Heulandite, Continued.

Dehydrated—

Free Apertures:	Unknown
Effect of Dehydration:	Structure changes at 215° to Heulandite "B", structure unknown
Location of H_2O Molecules:	Coordinated to calcium ions in the channels
Largest Molecule Adsorbed:	NH_3 if partially dehydrated at 130°
Kinetic Diameter, σ, A:	2.6

Table 22. Clinoptilolite, Table 2.24, Breck

Structure Group:	7		
Reference:	179—181		
Chemical Composition			
Typical Oxide Formula:	$(Na_2,K_2)O \cdot Al_2O_3 \cdot 10\ SiO_2 \cdot 8\ H_2O$		
Typical Unit Cell Contents:	$Na_6\,[(AlO_2)_6(SiO_2)_{30}] \cdot 24\ H_2O$		
Variations:	Ca, K, Mg also present; Na, K $\gg$ Ca		
	Si/Al, 4.25 to 5.25		
Crystallographic Data			
Symmetry:	Monoclinic	Density:	2.16 g/cc
Space Group:	I 2/m	Unit Cell Volume:	2100 A^3
Unit Cell Constants:	a = 7.41 A	X-Ray Powder Data:	Table 3.8
	b = 17.89 A		
	c = 15.85 A		
	$\beta = 91°29'$		
Structural Properties			
Framework:	Possibly related to heulandite but not determined		
Void volume:	0.34 cc/cc	Framework density:	1.71 g/cc
Dehydrated –			
Effect of Dehydration:	Very stable – in air to 700°C		
Largest Molecule Adsorbed:	O_2		
Kinetic Diameter, σ, A:	3.5		

Appendix III

X-RAY POWDER DIFFRACTION PATTERNS OF COMMON ZEOLITES
FOUND IN SEDIMENTARY ROCKS

Arthur J. Gude, 3rd

The following X-ray powder diffraction data are typical of those for
common zeolite minerals in sedimentary rocks. Data for potassium feldspar are included because this mineral is a common authigenic phase
associated with zeolites in such environments. With few exceptions, the
data were obtained using Ni-filtered Cu$K\alpha$ radiation. The absence of a
value for relative intensity indicates that the intensity of that line
is less than 1 on a scale of 1 to 100.

REFERENCES

Aumento, F. (1966) Thermal transformations of stilbite: Canadian J.
   Earth Sci. 3, 351-365.
Černy, P. and Povondra, P. (1965) Re-examination of two Moravian
   natrolites. Acta Universit. Carol. Geol., No. 2, 113-128.
Eberlein, G.D., Erd, R.C., Weber, F., and Beatty, L.B. (1971) New
   occurrence of yugawaralite from the Chena Hot Springs area,
   Alaska. Amer. Mineral. 56, 1699-1716.
Gude, A.J., 3rd and Sheppard, R.A. (1966) Silica-rich chabazite from
   the Barstow Formation, San Bernardino County, California. Amer.
   Mineral. 51, 909-915.
Iijima, A. and Harada, K. (1969) Authigenic zeolites in zeolitic
   palagonite tuffs on Oaho, Hawaii. Amer. Mineral. 54, 182-197.
Liou, J.G. (1971) P-t stabilities of laumontite, wairakite, lawsonite,
   and related minerals in the system $CaAl_2Si_2O_8-SiO_2-H_2O$. J.
   Petrology 12, 379-411.
Merkle, A.B. and Slaughter, M. (1968) Determination and refinement of
   the structure of heulandite. Amer. Mineral. 53, 1120-1138.
Seki, Yotaro (1968) Synthesized wairakite: their difference from
   natural wairakites. J. Geol. Soc. Japan 74, 457-458.
Sheppard, R.A. and Gude, A.J., 3rd (1969) Chemical composition and
   physical properties of the related zeolites offretite and erionite.
   Amer. Mineral. 54, 875-886.
Sheppard, R.A. and Gude, A.J., 3rd (1973) Zeolites and associated
   authigenic minerals in tuffaceous rocks of the Big Sandy Formation,
   Mohave County, Arizona. U.S. Geol. Surv. Prof. Paper 830, 36 pp.

ANALCIME (Table 1, p. 207)

$(Na_{12.49}K_{.28})(Al_{13.37}Fe_{.42}Si_{34.32}O_{96})\cdot 16.05H_2O$

Wikieup, Arizona; euhedral crystals, Lower Marker Tuff, Big Sandy Formation, sample SW-3-2A (Sheppard and Gude, 1973).

Cubic, a = 13.680(2) Å; V = 2560 Å³; Space group = $Ia3d$.

hkl	d(Å)	I/Io	hkl	d(Å)	I/Io
200	6.84		642	1.828	4
211	5.58	90	651	1.737	38
220	4.84	36	800	1.710	12
321	3.656	22	741	1.684	13
400	3.420	100	644	1.659	6
332	2.917	80	660	1.612	
422	2.792	16	750	1.590	12
431	2.683	34	842	1.493	4
521	2.498	30	761	1.475	7
440	2.418	20	664	1.458	2
532	2.219	18	930	1.442	
620	2.163	4	932	1.411	
541	2.111	5	844	1.396	
631	2.017	4	941	1.382	
543	1.935	5	860	1.368	
640	1.897	22	10-11	1.354	
552	1.862	15			

CHABAZITE (Table 14, p. 214)

$(Ca)_2(Al_4Si_8O_{24})\cdot 12H_2O$ (ideal)

North Table Mountain, Golden, Colorado; crystals in lava flow (Gude and Sheppard, 1966).

Rhombohedral, a = 13.786(2), c = 15.065(4) Å; V = 2479.5 Å³; Space group = $R\bar{3}m$.

hkl	d(Å)	I/Io	hkl	d(Å)	I/Io	hkl	d(Å)	I/Io
101	9.357	50	325	2.027		226	2.029	
110	6.893	10	207	2.025				
102	6.371	5	504	2.017				
201	5.550	9	600	1.990				
003	5.022	30	431	1.946	1			
202	4.678	6	217	1.943				
211	4.323	76	424	1.935	1			
113	4.059	1	520	1.912				
300	3.980	2	432	1.899	3			
212	3.871	28	505	1.871				
104	3.592	23	514	1.863				
220	3.447	13	108	1.860				
311	3.234	6	603	1.850	3			
204	3.185	5	416	1.8079	8			
303	3.119		611	1.8075				
312	3.031	2	425	1.8060	8			
401	2.928	100	317	1.8045				
105	2.921		208	1.7959				
214	2.894	30	523	1.7867				
223	2.842	3	612	1.7697	1			
402	2.775	4	515	1.7470	2			
321	2.695		407	1.7457				
205	2.690	7	434	1.7406	12			
410	2.605	10	218	1.7379				
322	2.574	2	440	1.7232	5			
006	2.511	5	336	1.6950				
215	2.506		531	1.6947	11			
314	2.487		701	1.6947				
116	2.359	2	327	1.6922				
501	2.358	2	009	1.6739				
404	2.339		532	1.6634	2			
413	2.313	4	702	1.6634	3			
330	2.298	4	435	1.6457				
502	2.276	1	621	1.6446	4			
421	2.231	1	614	1.6392	3			
315	2.229		318	1.6369				
324	2.215		443	1.6299				
422	2.161		119	1.6266				
306	2.123	2	622	1.6170	1			
511	2.123		507	1.5986				
405	2.120		408	1.5926	2			
107	2.118	2	710	1.5814				
333	2.089	6	606	1.5595				
512	2.062		615	1.5583	4			

CHABAZITE (silica-rich)

$(Ca_{.19}Mg_{.19}Na_{1.64}K_{.13})(Al_{2.46}Si_{9.51}O_{24})\cdot 10H_2O$

Rainbow Basin, Barstow, California; Barstow Formation, San Bernardino County (Gude and Sheppard, 1966).

Rhombohedral, a = 13.712(1), c = 14.882(2) Å; V = 2423.1 Å³; Space group = $R\bar{3}m$.

hkl	d(Å)	I/Io	hkl	d(Å)	I/Io
101	9.282		226	2.0095	
110	6.856	22	504	2.0018	2
102	6.305	8	207	2.0015	
201	5.515	32	600	1.9791	
003	4.961	38	431	1.9356	2
202	4.641	4	424	1.9216	
211	4.297	100	217	1.9213	2
113	4.019	5	520	1.9014	5
300	3.958	5	432	1.8883	
212	3.842	20	505	1.8564	
104	3.550	47	514	1.8503	
220	3.428	21	603	1.8382	3
311	3.216	10	108	1.8378	
204	3.153	2	611	1.7976	
303	3.094	1	425	1.7918	
312	3.012	1	416	1.7918	12
401	2.911	62	317	1.7862	
105	2.887	22	523	1.7755	2
214	2.864	34	208	1.7752	8
223	2.820	6	612	1.7595	
402	2.757	3	515	1.7336	
321	2.680	2	434	1.7286	
205	2.661	6	407	1.7285	
410	2.591	11	218	1.7185	
322	2.558	3	440	1.7140	8
215	2.481	12	531	1.6854	
006	2.480		701	1.6854	
314	2.466		336	1.6807	
501	2.345		327	1.6760	4
116	2.332		532	1.6539	
404	2.321		702	1.6539	6
413	2.297		009	1.6535	
330	2.285	1	621	1.6367	4
502	2.263	2	435	1.6324	
421	2.219	2	614	1.6282	
315	2.208	4	443	1.6200	
324	2.198		318	1.6197	
422	2.149	1	622	1.6078	
511	2.111		119	1.6074	
405	2.102	3	507	1.5840	
306	2.102		408	1.5763	2
107	2.093		710	1.5728	
333	2.076		615	1.5470	
512	2.050		606	1.5470	5
325	2.0096				

CLINOPTILOLITE (Table 22, p. 218)

$(Na_{3.02}K_{2.15}Mg_{.12}Ca_{.83})(Al_{6.38}Si_{29.70}O_{72})\cdot14.33H_2O$

Last Chance Canyon, Kern County, California; Ricardo Formation, Pliocene (Sheppard *et al.*, 1965).

Monoclinic, $a = 17.24(3)$, $b = 17.939(8)$, $c = 7.404(2)$ Å; $\beta = 113°40.7'$; $V = 2097.1$ Å³.

hkl	d(Å)	I/Io	hkl	d(Å)	I/Io
020	8.97	100	260	2.796	34
200	7.89	50	132	2.732	15
001	6.78	28	261	2.725	15
220	5.93	16	152	2.551	
311	5.100	35	261	2.464	
111	5.244	19	441	2.457	
310	5.050		712	2.346	
131	4.651		223	2.362	
401	4.229	20	603	2.208	
331	3.974	22	730	2.110	
221	3.915	100	172	2.093	
241	3.741	66	752	1.975	
202	3.698	38	841	1.931	
712	3.555	18	082	1.870	
311	3.461	25	840 }		
222	3.421	30	243 }	1.806	
002	3.391	27	2-101	1.7315	
241	3.123	23	951	1.6740	
332	3.070	18	10-21	1.6660	
331	3.038	14	554	1.6310	
351	2.974	48	2-102	1.6142	
242	2.854	11			

ERIONITE (Table 7, p. 210)

$(Na_{3.95}K_{2.42}Ca_{.33}Mg_{.76})(Al_{7.05}Si_{28.16}O_{72})\cdot22.6H_2O$

Tecopa, Inyo County, California; Tuff A, Pleistocene Lake Tecopa, Sample T4-61A (Sheppard and Gude, 1969).

Hexagonal, $a = 13.214(2)$, $c = 15.041(4)$ Å; $V = 2274.4$ Å³; Space group = $P6_3/mmc$.

hkl	d(Å)	I/Io	hkl	d(Å)	I/Io
100	11.44	100	314	2.425	
101	9.11	11	412	2.370	
002	7.52	7	305	2.362	3
110	6.61	73	116	2.344	3
102	6.28	5	323	2.376	
200	5.72	16	206	2.296	
201	5.35	14	500	2.289	3
112	4.964		404	2.277	
103	4.592	8	501	2.263	2
202	4.554	12	413	2.235	2
210	4.325	67	330	2.202	11
211	4.157	24	502	2.190	
300	3.815	37	315	2.183	
203	3.771		216	2.170	
004	3.760		420	2.164	
212	3.749	65	324	2.153	
301	3.698		421	2.141	65
104	3.572	24	332	2.114	2
302	3.402	4	107	2.112	6
220	3.304	39	306	2.095	3
213	3.275	25	503	2.082	2
114	3.268		213	2.080	3
310	3.174		414	2.078	5
204	3.142		422	2.073	1
311	3.106	12	405	2.073	
303	3.036	5	510	2.055	2
222	3.025	5	511	2.036	5
312	2.924	10	207	2.012	2
105	2.909	10	226	1.997	1
400	2.861	60	423	1.986	2
214	2.838	50	512	1.983	4
401	2.810	52	325	1.978	3
313	2.682		313	1.967	3
304	2.678		504	1.955	
402	2.674	15	217	1.924	
205	2.663	12	415	1.921	
320	2.625	8	600	1.907	1
321	2.586		513	1.902	
006	2.507		334	1.900	
410	2.497	20	601	1.892	
403	2.485		406	1.885	
224	2.482	17	430	1.881	6
322	2.479		008	1.880	2
215	2.470		424	1.875	
411	2.464		307	1.872	
106	2.449		431	1.867	3
			108	1.855	

FELDSPAR, POTASSIUM

$KAlSi_3O_8$ (Ideal)

Mud Hills, near Barstow, California; sample M4-2B, Skyline tuff (boron rich) (Sheppard and Gude, 1969).

Monoclinic, $a = 8.582(4)$, $b = 12.962(4)$, $c = 7.158(2)$ Å; $\beta = 116°2.3'$; $V = 715.3(4)$; Space group = $C2/m$.

hkl	d(Å)	I/Io	hkl	d(Å)	I/Io
110	6.627	8	223	2.222	
020	6.481	12	132	2.209	
111	5.847	12	330	2.195	
021	4.565		060	2.160	20
201	4.224	65	241	2.122	14
111	3.933	20	401	2.120	
200	3.855		402	2.112	
130	3.769	73	311	2.064	7
131	3.607	15	202	2.048	
221	3.539	13	061	2.008	13
112	3.449	47	422	1.9664	11
220	3.313	100	222	1.9492	
202	3.276	72	333	1.9277	11
040	3.240		400	1.9201	
002	3.216	95	351	1.9095	
131	2.924	51	403	1.8846	8
222	2.894		260	1.8823	
041	2.880	13	331	1.8477	
132	2.756		420	1.8459	
312	2.601	18	113	1.8326	
221	2.576	22	152	1.8316	6
241	2.571	30	423	1.8252	
112	2.541	11	350	1.8035	
310	2.521	10	262	1.7932	
240	2.481	5	062	1.7889	23
151	2.411	6	204	1.7879	
331	2.384	10	043	1.7879	
113	2.311	8			
332	2.262				

221

FERRIERITE (Table 20, p. 217)

$(Na_{1.8}K_{1.4}Mg_{.6})(Al_{4.4}Si_{31.6}O_{72})$·18H$_2$O

Agoura, California; crystals in brecciated porphyritic andesite (Wise et al., 1969).

Orthorhombic, a = 19.01(2), b = 14.135(7), c = 7.486(6) Å; V = 2011.9 Å³; Space group = Immm.

hkl	d(Å)	I/Io	hkl	d(Å)	I/Io
110	13.59	15	510	3.575	8
200	9.08	44	202	3.471	100
111	6.57	38	350	3.392	54
130	6.06	22	222	3.287	39
021	5.79	12	530	3.206	67
201	5.79	15	531	2.948	3
131	4.07	3	332	2.891	6
330	4.53	30	422	2.785	7
420	4.16	8	460	2.785	7
150	3.995	60	171	2.696	9
241	3.835	18	080	2.560	9
002	3.756	21	352	2.577	18

GISMONDINE (Table 5, p. 209)

$(Na_{1.60}K_{.80}Ca_{4.64})(Al_{13.52}Fe_{.80}Si_{18.40}O_{64})$·32.3H$_2$O

Round Top, Honolulu, Hawaii; spherulitic aggregates in vesicular palagonite glass (Iijima and Harada, 1969).

Monoclinic, a = 10.02(2), b = 10.63(2), c = 9.83(2) Å; β = 92°42'; Space group = P2$_1$/c.

hkl	d(Å)	I/Io	hkl	d(Å)	I/Io
110	7.285	56	322	2.498	32
110	7.285		322	2.413	15
111	5.941	10	033	2.405	17
111	5.779	10	332	2.374	22
002	4.916	59	240	2.345	
021	4.672	15	240	2.345	24
021	4.461	17	411	2.342	
012	4.174	71	313	2.331	12
121	4.054	37	142	2.285	12
211	3.450	24	402	2.193	12
211	3.435		323	2.103	15
221	3.428	17	242	2.101	
202	3.396		143	2.031	12
122	3.382	51	333	1.981	39
212	3.331	39	511	1.914	
221	3.277		115	1.913	15
003	3.269	34	115	1.884	
212	3.184	68	502	1.883	24
310	3.147		234	1.853	
131	3.131	61	512	1.850	12
013	3.065	10	324	1.822	
311	3.024	19	440	1.821	24
222	2.970	12	440	1.821	
023	2.789		512	1.801	
123	2.779	54	125	1.801	12
321	2.742		441	1.801	
312	2.723	100	215	1.779	
302	2.708	34	343	1.776	12
213	2.704		153	1.748	22
040	2.655	10	522	1.728	15
140	2.566		531	1.727	
140	2.566		225	1.708	12
041	2.563		433	1.708	

HARMOTOME (Table 4, p. 209)

$(Na_{1.46}K_{.65}Ba_{.78}Sr_{.01}Ca_{.16}Mg_{.51})$
$(Al_{4.04}Se_{.54}Si_{11.31}O_{32})$·11.2·H$_2$O

Wikieup, Arizona, Sample W7-44, acicular crystals (Sheppard and Gude, 1971).

Monoclinic, a = 9.921(4), b = 14.135(9), c = 8.685(6) Å; β = 124°55'; V = 998.6 Å³.

hkl	d(Å)	I/Io	hkl	d(Å)	I/Io
100	8.13	71	131	3.168	83
001	7.12	100	041	3.165	
011	6.36	68	312	3.142	
120	5.34	11	311	3.140	61
021	5.02	47	232	3.081	
201	4.960	12	230	3.079	19
102	4.285		321	2.941	31
101	4.280	39	032	2.930	
131	4.078	66	103	2.739	26
212	3.914		703	2.737	
210	3.909	19	102	2.726	
012	3.452	20	142	2.607	40
140	3.241	53	112	2.687	
301	3.220	38	221	2.675	48
			151	2.671	61
			051	2.627	
			323	2.535	
			320	2.532	23

HEULANDITE (Table 21, p. 217)

$(Ca_{3.5}Sr_{.4}K_{.9})(Al_{9.3}Si_{26.8}O_{72})\cdot 26.1H_2O$

Giedelsbach, Switzerland; crystals with fluorite (Merkle and Slaughter, 1968).

Monoclinic, a = 17.73, b = 17.82, c = 7.43 Å; β = 116°20'; Space group = Cm.

hkl	d(Å)	I/Io
020	8.909	80
200	7.945	70
001	6.659	60
220	5.930	10
311	5.256	50
310	5.077	70
131	4.639	60
401	4.369	20
421	3.923	100
241	3.726	20
321	3.565	20
222	3.428	70
002	3.329	20
422	3.177	50
510	3.129	40
350	2.957	90
530} 621}	2.802	70
532	2.735	40
042	2.667	10
152	2.531	20
261} 441} 712}	2.430	30
223	2.350	10
603	2.267	10
623	2.197	6
730	2.120	10
172	2.070	20
752} 753}	2.011	20

hkl	d(Å)	I/Io
841} 572}	1.963	30
082	1.851	10
840	1.814	10
102} 243}	1.771	30
2̄-101} 951}	1.722	10
0-101} 10̄-21}	1.698	20
004} 554}	1.665	10
024}	1.639	10
713} 2̄-102}	1.607	10
143} 154}	1.585	10
881} 882}	1.561	10
193} 11̄-31}	1.512	10
5̄-111} 10̄-63} 793}	1.473	10
0-121}	1.449	10
11̄-51}	1.431	6
10̄-60} 6̄-103}	1.401	10
11̄-05	1.360	10

LAUMONTITE (Table 6, p. 210)

$(Ca)(Al_2Si_4O_{12})\cdot 4H_2O$ (ideal)

Tanzawa Mountains, Kanagawa Prefecture, Japan (Liou, 1971).

Monoclinic, a = 14.737(4), b = 13.066(2), c = 7.550(3) Å; β = 111°58.2'; V = 1349 Å³; Space group = Cm.

hkl	d(Å)	I/Io
110	9.43	78
200	6.830	56
201	6.178	9
111	5.036	18
220	4.724	16
221	4.490	32
130	4.153	100
131	3.762	8
401	3.657	42
002	3.506	94
131	3.404	26
312	3.358	34
040	3.265	63
311	3.196	45
402	3.091	4
420	3.031	45
240	2.947	8
511} 422}	2.876	38
512	2.644	2

hkl	d(Å)	I/Io
331	2.628	3
241	2.571	34
132	2.537	3
222	2.517	5
203	2.515	9
601	2.452	7
441	2.437	43
403	2.388	2
151	2.358	23
350	2.267	9
622	2.215	8
060	2.178	8
620	2.150	28
332	2.087	4
532	1.990	5
601	1.957	19
731	1.887	2

MORDENITE (Table 18, p. 216)

$(Na_{1.79}K_{.37}Mg_{.24}Ca_{.82})$
$(Al_{4.05}Fe_{.15}Si_{19.78}O_{48})\cdot 11.75H_2O$

Mud Hills, Barstow, California; sample M4-54B, Skyline tuff, near base (Sheppard and Gude, 1969).

Orthorhombic, a = 18.15(2), b = 20.48(2), c = 7.512(9) Å; V = 2783.0 Å³; Space group = Cmcm.

hkl	d(Å)	I/Io	hkl	d(Å)	I/Io
110	11.34	3	330	3.781	65
200	9.51	50	510	3.672	12
020	7.07	38	040	3.534	100
011	6.61	3	202	3.483	10
310	5.78	15	240	3.312	35
220	5.67	14	312	3.142	12
400	4.75	2	431	3.055	12
130	4.56	1	402	2.941	11
031	3.987	35	051	2.645	7
420	3.943	35	350	2.582	10

223

NATROLITE (Table 15, p. 214)

$(Na_2)(Al_2Si_3O_{10}) \cdot 2H_2O$ (Ideal)

Palacov, Northeastern Moravia, Czechoslovakia (Černy and Povondra, 1965).

Orthorhombic, a = 18.284, b = 18.620, c = 6.592 Å; Space group = $Fdd2$.

hkl	d(Å)	I/Io	hkl	d(Å)	I/Io
220	6.49	100	800	2.288	6
111	5.90	65	280	2.256	6
040	4.65	40	731		
400	4.56	25	602	2.231	4
131	4.37	25	262	2.189	10
311	4.35	30	660	2.174	18
240	4.15	50	622		
420	4.10	25	430	2.071	4
331	3.63	4	840	2.048	6
440	3.20	10	313		
151	3.20	20	751	2.024	4
511	3.15	25	462		
022	3.10	14	333	1.960	4
292			991	1.907	4
260	2.945	18	513	1.872	6
222			391		
351	2.860	45	822	1.824	6
531	2.837	35	860		
460	2.569	10	533	1.797	10
422			771	1.793	16
171	2.441	12	482	1.754	4
711	2.406	18	842		
000	2.321	8	591	1.737	6
442					

OFFRETITE (Table 8, p. 211)

$(Na,K_2,Ca)_{4.5}(Al_9Si_{27}O_{72}) \cdot 27H_2O$ (Ideal)

Mount Simionse, Montbrison, Loire, France; type locality (Sheppard and Gude, 1969).

Hexagonal, a = 13.291(2), c = 7.582(6) Å; V = 1159.9 Å3; Space group = $P\bar{6}m2$.

hkl	d(Å)	I/Io	hkl	d(Å)	I/Io
100	11.51	100	411	2.384	
001	7.58	20	113	2.362	
110	6.64	35	203	2.314	
101	6.33		500	2.302	5
200	5.76		402	2.292	
111	4.998	4	330	2.215	22
201	4.584	59	501	2.203	
210	4.350	43	213	2.185	
300	3.837		420	2.175	2
002	3.792		322	2.167	
211	3.773	11	331	2.126	4
102	3.601	3	303	2.111	4
301	3.423	2	412	2.094	2
220	3.323	22	421	2.091	2
112	3.293		510	2.067	
310	3.192	17	223	2.012	
221	3.043		511	1.994	2
331	2.942	3	313	1.982	
400	2.880	64	502	1.968	2
212	2.858	15	600	1.918	
302	2.697		332	1.913	
401	2.690	3	403	1.899	
320	2.641	4	004	1.896	
003	2.528		430	1.892	
410	2.512		422	1.887	1
222	2.499	20	104	1.870	
104	2.494		601	1.860	
321	2.469		520	1.843	3
103	2.442		431	1.836	5
312					

PHILLIPSITE (Table 3, p. 208)

$(Na_{1.70}K_{1.54}Mg_{.78}Ca_{0.11})$
$(Al_{3.31}Fe_{.26}Si_{12.02}O_{32}) \cdot 9.62H_2O$

Wikieup, Arizona. Sample WT-25A, prismatic crystals, upper marker tuff (Sheppard and Gude, 1973).

Orthorhombic (pseudo-tiombic), a = 9.969(7), b = 14.139(7), c = 14.16(1) Å; V = 1992.1 Å3.

hkl	d(Å)	I/Io	hkl	d(Å)	I/Io
101	8.14	25	151	2.671	
002	7.06		242	2.669	29
020	7.07	71	313	2.665	
012	6.33		331	2.664	
121	5.34	30	125	2.542	10
022	5.00		323	2.533	
200	4.97		341	2.384	
103	4.26	43	026	2.239	
113	4.26	14	062	2.236	
131			305	2.154	
200	4.08		343	2.152	
220	4.07	28	315	2.129	
032	3.92		351	2.127	
123	3.652	5	326	2.042	
222	3.528		262	2.039	
014	3.434		117	1.963	
141	3.242		046		
301	3.229	64	155	1.962	64
024	3.166		064		
042	3.163		171	1.960	
133	3.162		424	1.956	
311	3.148	100	442	1.955	
321	2.937	35	361	1.904	
204	2.884		246	1.826	
240	2.831		264	1.825	
034	2.631		452	1.806	
105	2.724		008	1.770	
143	2.721	7	080	1.767	10
115	2.675	25			

STILBITE

$(Na_2Ca_4)(Al_{10}Si_{26}O_{72})\cdot28H_2O$ (Ideal)

Cape Blomindon, Nova Scotia; cruciform twins in sheaf-like aggregates in basalt (Aumento, 1966).

Monoclinic, a = 13.681, b = 18.128, c = 11.300 Å; β = 129°8'; Space group = C2/m.

hkl	d(Å)	I/Io	hkl	d(Å)	I/Io
001	8.77	100	3̄14 / 3̄51	2.791	33
020	9.06		404	2.739	5
200	5.306	6	441	2.586	
221	5.458		222	2.548	12
202	5.480		170	2.516	7
131	5.292	11	360	2.297	3
220	4.579	38	072	2.230	2
222	4.609		553	2.173	2
311	4.319		024	2.129	3
132	4.254	14	173	2.065	7
041	4.043		464	2.029	10
203	4.026	94	380	1.908	5
242	3.746	13	054	1.875 / 1.873	2
113	3.492	9	406	1.826	17
402	3.418		506	1.813	5
403	3.364	22	446	1.731	5
330	3.221	18	204	1.684	
152	3.052		575	1.673	3
243	3.017	71	394	1.640	3
241	2.837	6	045	1.635	1
052	2.793	3	466	1.591	8
			537	1.577	8

WAIRAKITE (Table 1, p. 208)

$(Ca)(Al_2Si_4O_{12})\cdot2H_2O$ (Ideal)

Natural mineral, Yugami, Japan (Seki, 1968).

Monoclinic (pseudo-cubic), a = 13.65, b = 13.66, c = 13.56 Å; β = 90°20'.

hkl	d(Å)	I/Io	hkl	d(Å)	I/Io
200	6.84	36	440	2.416	12
211	5.57	100	350	2.352	48
220	4.83	41	244	2.264	4
321	3.63	25	611	2.216	13
400	3.41	100	620	2.159	4
004	3.39	94	206	2.144	5
141	3.22	15	541	2.110	2
240	3.05	9	415	2.092	3
332	2.913	59	316	1.998	4
332	2.902	61	550	1.928	3
422	2.784	9		1.894	9
422	2.776	10	640	1.888	9
431	2.679	24	552	1.861	5
431	2.670	12	721	1.856	7
521	2.494	6	217	1.843	5
521	2.489	18	732	1.731	19

YUGAWARALITE

$(Na_{.02}K_{.02}Ca_{1.85}Sr_{.02})$
$(Al_{4.15}Fe_{.01}Si_{11.84}O_{32})\cdot16.4H_2O$

Chena Hot Springs area, Alaska; 1 mm lath-like crystals (Eberlein et al., 1971).

Monoclinic, a = 10.034(2), b = 13.997(3), c = 6.725(2) Å; β = 111°11'; v = 881.5 Å³; Space group = Pa.

hkl	d(Å)	I/Io	hkl	d(Å)	I/Io
010	14.0	6	202	3.192	5
110	7.78	5	211	3.150	
020	7.00	60	002	3.135	5
001	6.27	5	212	3.113	5
111	5.81	64	141	3.069	
	5.72		012	3.056	100
120	5.61	3	041	3.059	
121	4.72		310	3.047	
200	4.68	73	122	3.030	
021	4.67		321	2.994	
030	4.67		221	2.935	5
201	4.64		222	2.905	17
211	4.41	64	022	2.861	19
210	4.40	13	320	2.851	5
111	4.29	8	240	2.803	2
130	4.18	32	050	2.799	2
220	3.892	20	241	2.794	
221	3.868	8	141	2.765	
121	3.791	4	231	2.727	14
021	3.743		213	2.715	
040	3.499	3	133	2.701	12
230	3.305	5	051	2.582	
231	3.290	6	132	2.657	18
140	3.278		211	2.646	9
112	3.263				
131	3.243	7			
201	3.233				
311	3.223	30			

X

S